Lecture Notes in Computer Science 5818

Commenced Publication in 1973
Founding and Former Series Editors:
Gerhard Goos, Juris Hartmanis, and Jan van Leeuwen

María J. Blesa   Christian Blum
Luca Di Gaspero   Andrea Roli
Michael Sampels   Andrea Schaerf (Eds.)

Hybrid Metaheuristics

6th International Workshop, HM 2009
Udine, Italy, October 16-17, 2009
Proceedings

 Springer

Volume Editors

María J. Blesa
Christian Blum
Universitat Politècnica de Catalunya, LSI
Omega 213 Campus Nord, Jordi Girona 1-3, 08034 Barcelona, Spain
E-mail: {mjblesa, cblum}@lsi.upc.edu

Luca Di Gaspero
Andrea Schaerf
Università di Udine, DIEGM
Via delle Scienze 208, 33100 Udine, Italy
E-mail: {luca.digaspero, schaerf}@uniud.it

Andrea Roli
Università di Bologna, DEIS
Sede di Cesena, Via Venezia 52, 47521 Cesena, Italy
E-mail: andrea.roli@unibo.it

Michael Sampels
Université Libre de Bruxelles, IRIDIA
CP 194/6, Avenue Franklin D. Roosevelt 50, 1050 Bruxelles, Belgium
E-mail: msampels@ulb.ac.be

Library of Congress Control Number: 2009935813

CR Subject Classification (1998): I.2.8, F.2, G.1.6, F.1, G.2

LNCS Sublibrary: SL 1 – Theoretical Computer Science and General Issues

ISSN   0302-9743
ISBN-10   3-642-04917-6 Springer Berlin Heidelberg New York
ISBN-13   978-3-642-04917-0 Springer Berlin Heidelberg New York

springer.com

© Springer-Verlag Berlin Heidelberg 2009
Printed in Germany

Typesetting: Camera-ready by author, data conversion by Scientific Publishing Services, Chennai, India
Printed on acid-free paper   SPIN: 12772605   06/3180   5 4 3 2 1 0

Preface

The International Workshop on Hybrid Metaheuristics was established with the aim of providing researchers and scholars with a forum for discussing new ideas and research on metaheuristics and their integration with techniques typical of other fields. The papers accepted for the sixth workshop confirm that such a combination is indeed effective and that several research areas can be put together. Slowly but surely, this process has been promoting productive dialogue among researchers with different expertise and eroding barriers between research areas.

The papers in this volume give a representative sample of current research in hybrid metaheuristics. It is worth emphasizing that this year, a large number of papers demonstrated how metaheuristics can be integrated with integer linear programming and other operations research techniques. Constraint programming is also featured, which is a notable representative of artificial intelligence solving methods. Most of these papers are not only a proof of concept – which can be valuable by itself – but also show that the hybrid techniques presented tackle difficult and relevant problems.

In keeping with the tradition of this workshop, special care was exercised in the review process: out of 22 submissions received, 12 papers were selected on the basis of reviews by the Program Committee members and evaluations by the Program Chairs. Reviews were in great depth: reviewers sought to provide authors with constructive suggestions for improvement. Special thanks are extended to the Program Committee members who devoted their time and effort. Special gratitude is due to Andrea Lodi and Vittorio Maniezzo, who both accepted our invitation to give an overview talk.

The present selection of papers will be of interest not only to researchers working on optimization problems and constraint satisfaction problems by integrating metaheuristics with other areas. We hope that those who participated in HM 2009 will be assisted in making connections between their own specific research areas and others.

August 2009

María J. Blesa
Christian Blum
Luca Di Gaspero
Andrea Roli
Michael Sampels
Andrea Schaerf

Organization

General Chairs

Luca Di Gaspero   Università di Udine, Italy
Andrea Schaerf   Università di Udine, Italy

Program Chairs

María J. Blesa   Universitat Politècnica de Catalunya, Barcelona, Spain
Christian Blum   Universitat Politècnica de Catalunya, Barcelona, Spain
Andrea Roli   Università di Bologna, Italy
Michael Sampels   Université Libre de Bruxelles, Belgium

Program Committee

Mauro Birattari   Université Libre de Bruxelles, Belgium
Jürgen Branke   Universität Karlsruhe, Germany
Marco Chiarandini   Syddansk Universitet, Denmark
Carlos Cotta   Universidad de Málaga, Spain
Karl Dörner   Universität Hamburg, Germany
Andreas T. Ernst   CSIRO, Australia
Antonio J. Fernández   Universidad de Málaga, Spain
Paola Festa   Università di Napoli Federico II, Italy
José E. Gallardo   Universidad de Málaga, Spain
Thomas Jansen   University College Cork, Ireland
Joshua Knowles   University of Manchester, UK
Andrea Lodi   Università di Bologna, Italy
Manuel López-Ibáñez   Napier University, UK
Rafael Martí   Universitat de València, Spain
Daniel Merkle   Syddansk Universitet, Denmark
Bernd Meyer   Monash University, Australia
Martin Middendorf   Universität Leipzig, Germany
J. Marcos Moreno   Universidad de La Laguna, Spain
José A. Moreno   Universidad de La Laguna, Spain
Nysret Musliu   Technische Universität Wien, Austria
Boris Naujoks   Technische Universität Dortmund, Germany
David Pelta   Universidad de Granada, Spain
Steven Prestwich   4C, Cork, Ireland
Christian Prins   Université de Technologie de Troyes, France

Local Organization

Table of Contents

Hybrid Metaheuristic for the Assembly Line Worker Assignment and
Balancing Problem 1
   *Antonio Augusto Chaves, Luiz Antonio Nogueira Lorena, and
   Cristobal Miralles*

An ELSxPath Relinking Hybrid for the Periodic Location-Routing
Problem 15
   Caroline Prodhon

Hybridizing Beam-ACO with Constraint Programming for Single
Machine Job Scheduling 30
   *Dhananjay Thiruvady, Christian Blum, Bernd Meyer, and
   Andreas Ernst*

Multiple Variable Neighborhood Search Enriched with ILP Techniques
for the Periodic Vehicle Routing Problem with Time Windows 45
   Sandro Pirkwieser and Günther R. Raidl

A Hybridization of Electromagnetic-Like Mechanism and Great Deluge
for Examination Timetabling Problems 60
   Salwani Abdullah, Hamza Turabieh, and Barry McCollum

Iterative Relaxation-Based Heuristics for the Multiple-choice
Multidimensional Knapsack Problem 73
   Saïd Hanafi, Raïd Mansi, and Christophe Wilbaut

Solving a Video-Server Load Re-Balancing Problem by Mixed Integer
Programming and Hybrid Variable Neighborhood Search 84
   Jakob Walla, Mario Ruthmair, and Günther R. Raidl

Effective Hybrid Stochastic Local Search Algorithms for Biobjective
Permutation Flowshop Scheduling 100
   Jérémie Dubois-Lacoste, Manuel López-Ibáñez, and Thomas Stützle

Hierarchical Iterated Local Search for the Quadratic Assignment
Problem 115
   Mohamed Saifullah Hussin and Thomas Stützle

Incorporating Tabu Search Principles into ACO Algorithms 130
   Franco Arito and Guillermo Leguizamón

A Hybrid Solver for Large Neighborhood Search: Mixing Gecode and
EasyLocal++ 141
   Raffaele Cipriano, Luca Di Gaspero, and Agostino Dovier

Multi-neighborhood Local Search for the Patient Admission Problem ...   156
    Sara Ceschia and Andrea Schaerf

Matheuristics: Optimization, Simulation and Control   171
    Marco A. Boschetti, Vittorio Maniezzo, Matteo Roffilli, and
    Antonio Bolufé Röhler

Author Index...   179

Hybrid Metaheuristic for the Assembly Line Worker Assignment and Balancing Problem*

Antonio Augusto Chaves[1], Luiz Antonio Nogueira Lorena[1],
and Cristobal Miralles[2]

[1] Laboratory of Computing and Applied Mathematics,
National Institute for Space Research, São José dos Campos, Brazil
[2] ROGLE-Departamento Organización de Empresas,
Universidad Politécnica de Valencia, Valencia, Spain
{chaves,lorena}@lac.inpe.br, cmiralles@omp.upv.es

Abstract. The Assembly Line Worker Assignment and Balancing Problem (ALWABP) appears in real assembly lines which we have to assign given tasks to workers where there are some task-worker incompatibilities and considering that the operation time for each task is different depending upon who executes the task. This problem is typical for Sheltered Work Centers for the Disabled and it is well known to be NP-Hard. In this paper, the hybrid method Clustering Search (CS) is implemented to solve the ALWABP. The CS identifies promising regions of the search space by generating solutions with a metaheuristic, such as Iterated Local Search, and clustering them into clusters that are then explored further with local search heuristics. Computational results considering instances available in the literature are presented to demonstrate the efficacy of the CS.

1 Introduction

The World Health Organization estimates that 10% of global population, around 610 million people worldwide, is disabled. Of these, 386 million people are within the active labor age range, but experience very high unemployment rates.

Current practices for the treatment of physically and/or mentally handicapped individuals prescribe meaningful job activity as a means towards a more fulfilling life and societal integration [1]. In some countries active policies have been launched by national governments in order to achieve better labor integration of the disabled. These practices have facilitated the creation of Sheltered Work Centers for Disabled (referred to as SWD henceforth). These centers, which serve as a first work-environment for disabled workers, should be characterized by an environment where these workers can gradually get adapted to a working routine and developed their personal skills, before being fully integrated into the conventional labor market.

---

* The authors acknowledge FAPESP (process 2008/09242-9) for the financial support given to the development of this work.

M.J. Blesa et al. (Eds.): HM 2009, LNCS 5818, pp. 1–14, 2009.

This model of socio-labor integration tries to move away from the traditional stereotype that considers disabled people as unable to develop continuous professional work. Just as in any other firm, the SWDs compete in the market and must be flexible and efficient enough to adapt to market fluctuations and changes, the only difference being that the SWD is a Not-For-Profit organization. Thus, the potential benefits that may be obtained from increased efficiency usually implies growth of the SWD. This means more jobs for the disabled and the gradual integration of people with higher levels of disability; which are in fact the primary aims of the SWD.

Miralles et al. [2] revealed how in these centers the adoption of assembly lines provides many advantages, as the traditional division of work into single tasks can become a perfect tool for making certain worker disabilities invisible. In fact, an appropriate task assignment can even become a good therapeutic method for the rehabilitation of certain disabilities. But some specific constraints relative to time variability arise in this centers, in which case the balancing procedures applied in a SWD should be able to reconcile the following objectives:

1) to maximize the efficiency of the line by balancing the workload assigned to each available worker in each workstation;
2) to satisfy and respect the existent constraints in this environment due the human factors when assigning tasks to workers.

After analyzing some SWDs, Miralles et al. [2] observed some characteristics that can be found in this environment, which were the motivation for defining the Assembly Line Worker Assignment and Balancing Problem (ALWABP). This problem is known to be NP-hard and has a important application in the SWDs.

In general, an assembly line consists of a set tasks, each having an operation time, and a set of precedence constraints, usually represented by a precedence graph $G = (V, A)$ in which V is the set of tasks and an arc (i, j) is in A if task i must be executed before j and there is no task k that must be executed after i and before j. The ALWABP consists of assigning tasks to workstations, which are arranged in a predefined order, so that the precedence constraints are satisfied and some give measure of effectiveness is optimized. Therefore, in the SWD some workers can be very slow, or even incapable, when executing some tasks, but very efficient when executing some other tasks. So, the balancing of the line consists of a double assignment: (1) tasks to workstations; and (2) workers to workstations. Always respecting the incompatibilities among tasks and workers.

The main characteristics of the ALWABP have been listed by Miralles et al. [2] as follows:

a) a single product is assembled on the line;
b) task operation times and precedence constraints are known deterministically;
c) there is a serial line layout with k workstations;
d) k workers are available and the operation times of a task depends on the workers executing it;
e) each worker is assigned to only one workstation;

f) each task is assigned to only one workstation, provided that the worker selected for that workstation is capable of performing the task, and that the precedence constraints are satisfied.

When we aim to minimize the number of workstations, the problem is called ALWABP-1; and when the objective is to minimize the cycle time given a set of workstations, the problem is called ALWABP-2, the latter situation being more common in SWDs.

This paper presents an application of the hybrid method Clustering Search (CS) [3,4] to solve the ALWABP-2. The CS consists of detecting promising areas of the search space using a metaheuristic that generates solutions to be clustered. These promising areas should be exploited with local search heuristics as soon as they are discovered. The Iterated Local Search (ILS) [5] was the metaheuristic chosen to generate solutions for the clustering process. In this paper some improvements to the CS are also proposed.

The remainder of the paper is organized as follows. Section 2 reviews previous works about ALWABP. Section 3 describes the CS method and section 4 presents the CS applied to the ALWABP. Section 5 presents the computational results. Conclusions are reported in section 6.

2   Literature Review

The ALWABP was recently introduced by Miralles et al. [2]. The authors presented a mathematical model for ALWABP and a case study based on a Spanish SWD. Miralles et al. [6] proposed a basic branch and bound procedure with three possible search strategies and different parameters for solving the ALWABP.

Chaves et al. [7,8] proposed the use of hybrid method CS for solving the ALWABP, which was implemented using Simulated Annealing (SA) [9] to generate solutions to the clustering process.

There are some other problems with double assignment of tasks and resources to workstations. For example, some cost-oriented models assume that different equipment sets, each with a different cost, can be assigned to a workstation. In this sense, total cost must be minimized by optimally integrating design (selecting the machine type to locate at each activated workstation) and operating issues (assigning tasks to observe precedence constraints and cycle time restrictions). When these decisions are connected, the terms Assembly Line Design Problem (ALDP) [10] or Assembly System Design Problem (ASDP) [11] are frequently used in the literature.

Although ALWABP can be classified with problems of these types, it is not a cost-oriented problem in which there are alternative machines with different associated costs and the total cost has to be minimized. Furthermore, in ALWABP the available resources are constrained: there are unique workers each which can be assigned only once. In some cases workers have similar characteristics; but, even in these cases, an infinite number of workers is not available, as assumed in most ASDP problems.

Variants of the ALWABP have been studied in the literature: Miralles et al. [12] analyze the design of U-shaped assembly lines in the context found at SWD while Costa and Miralles [13] proposes models and algorithms to plan job rotation schedules in SWD.

3   Clustering Search Algorithm

The Clustering Search (CS) [3] is a hybrid method that aims to combine metaheuristics and local search heuristics, in which the search is intensified only in areas of the search space that deserve special attention (promising regions). The CS introduces an intelligence and priority to the choice of solutions to apply local search, instead of choosing randomly or apply local search in all solutions. Therefore, it is expected an improvement in the convergence process associated with a decrease in computational effort as a consequence of a more rational employment of the heuristics.

In fact in this paper the CS method is improved making it more efficient, robust and user-friendly. This is done through the implementation of a random perturbation and a control of the efficiency of local search heuristics. Moreover, the functions of each component of CS was defined more clearly, redesigning the structure of the CS. These improvements change the CS into a method easier to learn and easier to use.

The CS attempts to divide the search space and locate promising search regions by framing them in clusters. For the purposes of the CS, a cluster can be defined by three attributes $C = (c, v, r)$. The center c_i is a solution that represents cluster C_i, identifying its location within the search space. Instead of keeping all solutions generated by the metaheuristic and grouping them into clusters, just a part of these solutions characteristics are inserted to the center. The volume v_i is the number of solutions grouped into the cluster C_i. This volume determines when a cluster becomes promising. The inefficacy rate r_i is a control variable that verifies if the local search is improving the center c_i. The value of r_i indicates the number of consecutive times that the local search was applied to the cluster C_i and did not improve the solution. This attribute avoids that the local search is being performed by more than r_{max} times in bad regions or regions that have already been exploited by the heuristic.

In order to group similar solutions in clusters, the CS needs some form of distance measuring between two solutions. Therefore, for two solutions s_1 and s_2, the distance function $d(s_1, s_2)$ is defined as the number of locations in which s_1 and s_2 differ. For example, the distance of $s_1 = \{1010110\}$ and $s_2 = \{1100100\}$ is three $(d(s_1, s_2) = 3)$.

Initially, we need to define the number of clusters $(|C|)$. And, the cluster centers should be generated in order to represent different regions of search space. This paper uses a greed method based on maximum diversity to create the initial centers, which generate a large set with n $(n >> |C|)$ random solutions and a subset is selected with $|C|$ solutions that have the longest distance among themselves. In this method, the first solution is randomly chosen. The second solution must be the solution that offers the greatest distance for the first one. From the third solution, it is necessary to calculate the sum of distances among each candidate solution and the selected solutions. Then, the solution that has the biggest sum is also selected to be the initial center of one cluster.

The CS is an iterative method which has three main components: a meta-heuristic, a clustering process and a local search method. The hybrid strategy of the CS can be described by the flowchart illustrated in figure 1.

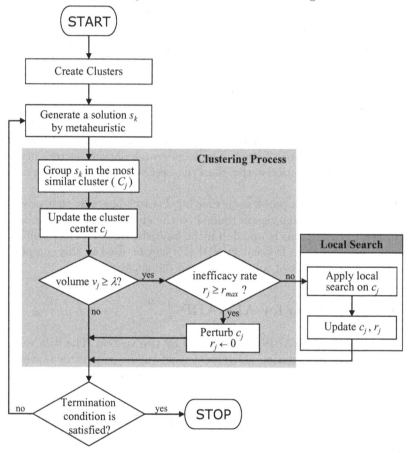

Fig. 1. Flowchart of the CS

A metaheuristic works as a solution generator to the clustering process. The algorithm performs independently of the others components and must be able to provide a great number of different solutions for enabling a broad analysis of the search space.

In each iteration of CS, a solution s_k is generated by a metaheuristic and sent to the clustering process. This solution is grouped in the most similar cluster

C_j, that is the cluster with the smallest distance d between the center c_j and the solution s_k. The clustering process aims to direct the search for supposedly promising regions.

The insertion of a new solution into a cluster should cause a perturbation in its center. This perturbation is called assimilation and consists of updating the center c_j with attributes of solution s_k. This process is the Path-Relinking method [14], which generates several solutions along the path that connects the center c_j and the solution s_k. This process is responsible for intensify and diversify the search into the cluster, for the reason that the new center is the best-evaluated solution obtained along the path, even if it is worse than the current center.

Then, the volume v_j is analyzed. If it has reached the threshold λ, this cluster may be a promising region. And, if the local search has achieved success in recent r_{max} applications in this promising cluster ($r_j < r_{max}$), the center c_j is better investigated with local search to accelerate the convergence process. Otherwise, if $r_j \geq r_{max}$ a random perturbation is performed in c_j, allowing the center go to another region of search space. The local search obtains success in a cluster j when one finds a solution better than the best solution obtained so far in this cluster (c_j^*).

The local search is a problem-specific local search heuristic that provides the exploitation of promising region framed by the cluster. The heuristic is applied on the center c_j and this is updated if the heuristic finds a better solution. The Variable Neighborhood Descent (VND) [15] can be used in this component, to analyze a large number of solutions around the cluster.

4   CS Algorithm for ALWABP

A solution of the ALWABP is represented by two vectors. The first vector represents the task/workstation assignments and the second vector represents the worker/workstation assignments. Figure 2 shows an example of one solution with 11 tasks, 5 workers and 5 workstations.

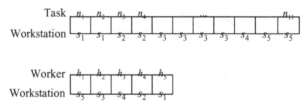

Fig. 2. An example of the solution representation

In the particular case of the CS developed for the ALWABP, we make use of penalties in the objective function. Let C_{time} be the cycle time and f_p and f_t be the infactibilities measures of the precedence constraints violations and the infeasible task/worker assignments; ω and δ be the multipliers for the values f_p and f_t. The objective function of ALWABP is defined as follows:

$$f(s) = C_{time} + (\omega * f_p + \delta * f_t) \tag{1}$$

The distance of two solutions (d) is the number of tasks assigned to different workstations between them. So, the distance increases when there is a large number of allocations to different workstations between the solutions.

4.1 Iterated Local Search

The CS uses the Iterated Local Search (ILS) [5] to generate solutions to the clustering process. ILS consists in the iterative application of a local search procedure to starting solutions that are obtained by the previous local optimum through a solution perturbation.

To apply an ILS algorithm, four components have to be specified: a procedure that generates an initial solution s_0, a perturbation that modifies the current solution $\hat{s}$ leading to some intermediate solution s', a local search procedure that returns an improved solution $\hat{s}'$, and an acceptance criterion that decides to which solution the next perturbation is applied.

In this paper we propose a method to generate an initial solution without violating the precedence network. The method is based in the Kilbridge-Wester heuristic [16] and it considers the number of tasks that precede each task. A ordered list is built, in ascending order, with tasks that have the smallest number of predecessors. If two or more tasks have the same number of predecessors, the order is chosen randomly. Then, we select a task of the list (starting from first up to last) and assigned it on an available workstation. The first tasks should be assigned to the first workstations, and so on. Thus, the workstations have the same number of assigned tasks. Finally, workers are randomly assigned to workstations. Therefore, the initial solution probably have only infeasibilities between tasks and workers.

In order to escape from local optima and to explore new regions of the search space, ILS applies perturbations to the current solution. A crucial issue concerns the strength of the perturbation. If it is too strong, ILS may behave like a random restart resulting in a very low probability of finding better solutions. On the other hand, if the perturbation is not strong enough, the local search procedure will rapidly go back to a previous local optimum.

We defined three types of perturbation movements that are randomly applied in our ILS procedure. The first consists in swapping two tasks between two workstations. The second moves one task from one workstation to another. And the third is obtained by swapping two workers between two workstation.

The strength of the perturbation is the number of solution components (tasks and workers) that are modified, and it is randomly defined in each iteration. A percentage β ($\beta \in [0.25, 0.75]$) of the number of tasks are altered by the perturbation.

We select the *Swap Worker* and *Swap Task* heuristics as the local search of ILS. This heuristics performs a descent from a solution s' until it reaches a local minimum ($\hat{s}'$). Further details of these methods are presented in section 4.3.

The acceptance criterion is biased towards diversification of the search since the best solution resulting from the local search phase is accepted if it improves the local minimum encountered so far or with a probability of 5%.

The condition used to stop the algorithm was the maximal number of iterations, which was defined as 5000 iterations. This is the termination condition of the CS.

4.2   Clustering Process

At each iteration of CS, the solution $\hat{s}'$ is grouped into the closest cluster C_j; that is, the cluster that minimizes the distance between the solution and the cluster center. The volume v_j is increased in one unit and the center c_j should be updated with new attributes of the solution $\hat{s}'$ (assimilation process).

The assimilation process uses the path-relinking method. The procedure starts by computing the symmetric difference between the center c_j and the solution $\hat{s}'$ $\Delta(c_j, \hat{s}')$; i.e., the set of moves needed to reach $\hat{s}'$ from c_j. A path of solutions is generated, linking c_j and $\hat{s}'$. At each step, the procedure examines all moves $m \in \Delta(c_j, \hat{s}')$ from the current solution s and selects the one that results in the best-cost solution, applying the best move to solution s. The set of available moves is updated. The procedure terminates when 30% of the solutions in the path have been analyzed, this is to stop the center from moving too far. The new center c_j is the best solution along this path. In this paper, one move is to swap the workstation in which one task of c_j is assigned by the workstation of this task in the $\hat{s}'$.

After performing the path-relinking, we must conduct an analysis of the volume v_j, verifying if this cluster can be considered promising. A cluster becomes promising when its volume reaches the threshold λ ($v_j \geq \lambda$). The value of λ was set equal to 10.

Then, if the volume v_j reached λ and the local search has been obtained success in this cluster ($r_j < r_{max}; r_{max} = 5$), an exploitation is applied in center c_j by local search heuristics. Otherwise, if the inefficacy rate r_j is greater than 5, we must apply a random perturbation to the center, swapping 30% of assigments task/workstation.

4.3   Local Search

The VND method is implemented as local search of CS, intensifying the search in neighborhood of a promising cluster C_j. Three types of moves are relevant in a VND for ALWABP: swap tasks, shift tasks and swap workers. So, our VND procedure uses three heuristics based on these moves, which seek to improve the center of a promising cluster.

The VND improvement methods are:

- *Swap Tasks*: to perform the best move, swapping two tasks that have been assigned to different workstations;
- *Shift Task*: to perform the best move, removing one task from a workstation and assigning it to another;
- *Swap Workers*: to perform the best move, swapping the workstation assignments of two workers.

---

algorithm CS

---

create the initial clusters of CS
{ metaheuristic – ILS}
generate an initial solution s_0

$\hat{s} \leftarrow$ LocalSearch (s_0)

while termination condition not satisfied **do**
    $s' \leftarrow$ Perturbation ($\hat{s}$)
    $\hat{s}' \leftarrow$ LocalSearch (s')
    $\hat{s} \leftarrow$ AcceptanceCriterion ($\hat{s}, \hat{s}'$)
    { clustering process }
    find the most similar cluster C_j to the solution $\hat{s}'$
    insert $\hat{s}'$ into C_j ($v_j \leftarrow v_j + 1$)
    update the center c_j ($c_j \leftarrow$ PR ($c_j, \hat{s}'$))
    if $v_j \geq \lambda$ **then**
        $v_j \leftarrow 0$
        if $r_j \geq r_{max}$ **then**
            apply a random perturbation in c_j
            $r_j \leftarrow 0$
        else
            { local search }
            $\hat{c}_j \leftarrow$ VND (c_j)
            if $f(\hat{c}_j) < f(c_j)$ **then**
                $c_j \leftarrow \hat{c}_j$
            end if
            if $f(\hat{c}_j) < f(c_j^*)$ **then**
                $r_j \leftarrow 0$
                $c_j^* \leftarrow \hat{c}_j$
            else
                $r_j \leftarrow r_j + 1$
            end if
        end if
    end if
end while

Fig. 3. CS algorithm for the ALWABP

If some heuristic obtains a better solution, VND returns to the first heuristic and the search continues from the better solution; otherwise, it changes to the next heuristic. The VND stopping condition is that there are no more improvements to be made to the incumbent solution. The center c_j is updated if the new solution is better than the previous one. On the other hand, the inefficacy rate r_j is reset if and only if the VND improves the best center found so far in this cluster (c_j^*). Otherwise, r_j is increased in one unit.

Figure 3 presents the CS pseudo-code.

5   Computational Results

The CS algortithm was coded in C++ and computational tests were executed on a 2.6 GHz Pentium 4 with 1 GB RAM. In order to test the CS, we have used the problem sets proposed by Chaves et al. [7] and available in the literature. These sets are composed of four families: Roszieg, Heskia, Wee-Mag and Tonge. The ALWABP benchmark is composed of 320 instances (80 in each family), enabling to extract conclusions about the overall behavior of CS against different kind of problem. The characteristics for each problem set (number of tasks (|N|), number of workers (|W|) and the order strength (OS) of the precedence network) are listed in table 1. The OS measures the relative number of precedence relations. That is, problems with a large strength are basically expected to be more complex than such with small OS values [17].

Table 1. ALWABP: Instances characteristics

Family	|N|	|W|	OS
Roszieg	25	4 (group 1-4) or 6 (group 5-8)	71,67
Heskia	28	4 (group 1-4) or 7 (group 5-8)	22,49
Tonge	70	10 (group 1-4) or 17 (group 5-8)	59,42
Wee-Mag	75	11 (group 1-4) or 19 (group 5-8)	22,67

Two exact approaches were tested in Chaves et al. [7,8]: a branch and bound algorithm proposed by Miralles et al. [6] and the commercial solver CPLEX 10.1 [18]. The results of CPLEX were better than the results of branch and bound algorithm. CPLEX found the optimal solutions only for instances of the smaller Roszieg and Heskia families and had memory overload in instances of the Tonge e Wee-Mag families, finding very poor feasible sub-optimal solutions. Sometimes CPLEX failed in finding feasible solutions even after several hours of computational time for the larger instances. Chaves et al. [7,8] also present the results of a version of the CS algorithm for ALWABP, and we use these results for comparison. The tests were also executed on a 2.6 GHz Pentium 4 with 1 GB RAM.

Tables 2 - 5 present the results for each set of instances. In the tables, the results are averaged in each line for each group of 10 instances with same

Table 2. ALWABP: Results for the Roszieg family of instances

Family	Group	best-known	CS best	CS avrg	CS t_b(s)	CS t(s)	ILS best	ILS avrg	ILS t(s)	Chaves et al. [7,8] best	Chaves et al. [7,8] avrg	Chaves et al. [7,8] t_b(s)	Chaves et al. [7,8] t(s)
Roszieg	1	20.1	20.1	20.2	0.8	3.8	20.1	20.4	3.4	20.1	20.2	2.2	5.2
	2	31.5	31.5	32.5	0.9	3.7	32.4	38.9	3.2	31.5	34.3	2.0	5.1
	3	28.1	28.1	28.5	0.6	3.8	28.2	28.7	3.3	28.1	28.1	2.0	5.2
	4	28.0	28.0	28.0	0.2	3.8	28.0	28.1	3.3	28.0	28.1	1.9	5.2
	5	9.7	9.7	10.7	1.3	5.2	10.3	11.8	4.6	9.7	10.2	3.5	6.0
	6	11.0	11.0	12.1	1.4	5.2	11.5	14.3	4.5	11.0	11.9	3.6	6.0
	7	16.0	16.0	16.9	1.5	5.2	16.5	18.7	4.6	16.0	16.2	3.5	6.0
	8	15.1	15.1	15.6	1.9	5.2	15.3	17.7	4.6	15.1	15.4	3.4	6.0
	average	19.94	19.94	20.57	1.09	4.48	20.29	22.32	3.94	19.94	20.57	2.75	5.59

Table 3. ALWABP: Results for the Heskia family of instances

Family	Group	best-known	CS best	CS avrg	CS t_b(s)	CS t(s)	ILS best	ILS avrg	ILS t(s)	Chaves et al. [7,8] best	Chaves et al. [7,8] avrg	Chaves et al. [7,8] t_b(s)	Chaves et al. [7,8] t(s)
Heskia	1	102.3	102.3	102.8	1.3	5.0	102.3	103.0	6.3	102.3	103.5	3.026	5.802
	2	122.6	122.6	123.8	1.4	5.0	122.7	124.2	6.3	122.6	123.7	2.482	5.736
	3	172.5	172.5	175.5	1.7	5.0	172.6	176.4	6.4	172.5	173.1	2.62	5.779
	4	171.2	171.2	171.7	1.4	5.1	171.3	171.8	6.4	171.2	171.8	2.668	5.789
	5	34.9	34.9	37.8	4.4	9.0	35.3	38.6	7.5	34.9	36.4	4.584	7.448
	6	42.6	42.6	44.7	3.4	9.0	43.6	45.7	7.5	42.6	44.3	4.232	7.386
	7	75.2	75.2	77.7	2.9	9.0	76.7	78.6	7.5	75.2	76.4	3.554	7.361
	8	67.2	67.2	70.7	3.6	9.0	68.1	72.4	7.5	67.2	70.2	4.054	7.376
	average	98.56	98.56	100.59	2.51	7.00	99.08	101.35	6.92	98.56	99.92	3.40	6.58

characteristics. The tables list for each method the best solution found (*best*), the average solution found out of the 20 runs (*avrg*) and the total computational time (t) in seconds. For our CS method and CS proposed by Chaves et al. [7,8], the tables also list the time in which the best solution was found (t_b).

The results in the tables show the efficacy of the proposed CS. We can observe that our CS and the method proposed by Chaves et al. [7,8] found the best-known solutions for all instances of the Roszieg and Heskia families, and these are optimal solutions as proved by the solver CPLEX. However, our CS finds the best solution in shorter time. These instances are easy to solve, but the tests are important to show that CS is able to obtain good solutions in terms of quality, as a result of the comparison with the optimal solutions.

For the Tonge family, the proposed method found best-known solutions in 76 of 80 instances. The solutions are better than the previously best-known solution in 73 tested instances. It is interesting to note that the solution found by our CS are about 29% better than the best solutions obtained by Chaves et al. [7,8]. Our method also were better in terms of average solutions (590% better), mainly, beacuse the method of Chaves et al. [7,8] found infeasible solutions in some instances.

Table 4. ALWABP: Results for the Tonge family of instances

Family	Group	best-known	CS				ILS			Chaves et al. [7,8]			
			best	avrg	t_b(s)	t(s)	best	avrg	t(s)	best	avrg	t_b(s)	t(s)
Tonge	1	96.7	96.7	116.6	64.0	122.2	120.0	135.3	102.5	107.5	732.8	40.6	58.0
	2	116.0	116.0	141.8	64.6	122.6	151.8	174.8	101.7	141.8	826.0	41.8	59.3
	3	167.1	167.7	199.4	66.2	122.8	214.6	236.5	102.6	179.5	778.7	39.4	58.0
	4	174.0	174.0	206.0	65.7	123.0	220.7	244.3	102.6	206.4	755.2	38.1	57.9
	5	41.3	41.3	51.3	101.1	183.0	64.2	71.1	151.1	71.9	877.7	55.7	83.9
	6	48.5	48.5	61.6	105.3	184.7	74.2	87.4	151.4	83.9	913.2	56.3	86.0
	7	77.8	77.8	93.0	100.1	184.5	113.7	129.3	151.4	132.6	912.0	53.6	84.9
	8	77.5	77.9	95.6	100.3	184.9	116.1	131.3	151.5	113.8	875.9	59.1	84.9
	average	99.86	99.99	120.64	83.42	153.48	134.41	151.25	126.86	129.68	833.93	48.06	71.61

Table 5. ALWABP: Results for the Wee-Mag family of instances

Family	Group	best-known	CS				ILS			Chaves et al. [7,8]			
			best	avrg	t_b(s)	t(s)	best	avrg	t(s)	best	avrg	t_b(s)	t(s)
Wee-Mag	1	29.0	29.0	32.7	94.3	163.2	31.2	35.5	51.0	33.6	59.0	52.8	69.5
	2	34.6	34.6	38.4	91.4	160.8	37.4	41.0	51.2	38.5	45.9	59.0	71.3
	3	50.8	50.8	56.7	96.0	160.4	54.3	61.3	51.0	56.2	67.1	57.3	69.9
	4	49.6	49.6	55.6	103.9	158.8	52.7	58.9	51.2	55.2	73.5	53.1	69.8
	5	13.1	13.1	20.9	141.2	248.6	16.7	45.4	64.8	19.4	504.2	59.0	99.4
	6	14.6	14.6	18.2	155.2	249.0	18.7	23.7	65.0	21.8	544.9	66.8	102.2
	7	21.2	21.2	27.1	148.0	244.8	25.1	34.1	64.6	30.6	370.0	69.7	100.9
	8	21.6	21.6	26.8	140.6	243.4	24.9	33.7	64.7	27.9	443.6	68.9	101.2
	average	29.31	29.31	34.56	121.31	203.65	32.63	41.70	57.92	35.40	263.51	60.82	85.52

Finally, for the last and larger class of instances (Wee-Mag), our CS found the best-known solutions for all instances and improved the best-known solution so far in 77 instances. The improvement of our CS in terms of quality of the solutions was about 20%. Again, the average solutions of our CS were also much better than the results of Chaves et al. [7,8] (660% better).

We can observe that ILS without the clustering process gave solutions in poorer quality than CS in all problem families. These results show the importance of the clustering process and the local search method for the CS. Indeed, the best solution is always found first by path-relinking or VND.

The proposed method was robust, producing average solutions close to best solutions. The computational times of our CS were competitive, finding good solutions in few seconds for the smaller instances and in a reasonable time for the larger instances. However, the improvements proposed for the CS became it slower than the CS proposed by Chaves et al. [7,8], this can be observed by analyzing of the computational time for the Tonge and Wee-Mag families.

The convergence of CS had an interesting behavior, allowing to find good solution in 25% of total computational time for the Roszieg family (best case) and in 60% of the total computational time for the Wee-Mag family (worst case). The features and parameters settings of CS are implemented in order to speed up the optimization process and avoid premature convergence to local optimum.

6   Conclusions

This paper presents a solution for the Assembly Line Worker Assignment and Balancing Problem (ALWABP) using the Clustering Search (CS) algorithm. The CS has been applied with success in some combinatorial optimization problems, such as the pattern sequencing [3], the prize collecting traveling salesman [19], the flowshop scheduling [20], the capacitated p-median [4], and others.

The idea of the CS is to avoid applying a local search heuristic to all solutions generated by a metaheuristic. The CS detects the promising regions in the search space and applies local search only in these regions. Then, to detect promising regions becomes an interesting alternative, preventing the indiscriminate application of the heuristics.

This paper reports results of different classes of instances to the ALWABP found by the CS. The CS got the best-known solution in 314 of 320 instances, and it defined new best solutions for 306 instances. The results show that our CS approach is competitive for solving the ALWABP.

We proposed some improvements for the CS, making it more efficient, robust and user-friendly. The results found by this new version are better than the CS proposed by Chaves et al. [7,8], in terms of quality of solution and in terms of robustness of the method.

In practical sense, the short computational times achieved enable rapid balancing of the assembly line. According to Miralles et al. [6], this is very important if we take into account that, due to the high absenteeism and the periodic physical and psychological checking of workers, the SWD manager knows only at the beginning of every working day which workers are available. Therefore approaches like CS, which provide good solutions in short computational times, are very desirable for the aids to set up an assembly line early in the day.

Apart from implementation in industrial software, further studies can be done analyzing other metaheuristics to generate solutions for the CS clustering process (e.g., Ant Colony System, Tabu Search and Genetic Algorithm), and also exploring other problems to which CS could be applied.

References

1. Bellamy, G.T., Horner, R.H., Inman, D.P.: Vocational habilitation of severely retarded adults: a direct service technology. University press, Baltimore (1979)
2. Miralles, C., García-Sabater, J., Andrés, C., Cardós, M.: Advantages of assembly lines in sheltered work centres for disabled: a case study. International Journal of Production Economics 110(1-2), 187–197 (2007)
3. Oliveira, A.C.M., Lorena, L.A.N.: Hybrid evolutionary algorithms and clustering search. In: Grosan, C., Abraham, A., Ishibuchi, H. (eds.) Hybrid Evolutionary Systems - Studies in Computational Intelligence, pp. 81–102. Springer, Heidelberg (2007)
4. Chaves, A.A., Lorena, L.A.N., Corrêa, F.A.: Clustering search heuristic for the capacitated p-median problem. Advances in Soft Computing 44, 136–143 (2008)

5. Lourenço, H., Martin, O., Stützle, T.: Iterated local search. In: Glover, F., Kochenberger, G.A. (eds.) Handbook of Metaheuristics. International Series in Operations Research & Management Science, vol. 57, pp. 320–353. Springer, New York (2003)
6. Miralles, C., García-Sabater, J.P., Andrés, C., Cardós, M.: Branch and bound procedures for solving the assembly line worker assignment and balancing problem: application to sheltered work centres for disabled. Discrete Applied Mathematics 156(3), 352–367 (2008)
7. Chaves, A.A., Miralles, C., Lorena, L.A.N.: Clustering search approach for the assembly line worker assignment and balancing problem. In: Proceedings of 37th International Conference on Computers and Industrial Engineering, Alexandria, Egypt, pp. 1469–1478 (2007)
8. Chaves, A.A., Miralles, C., Lorena, L.A.N.: Uma meta-heurística híbrida aplicada ao problema de balanceamento e designação de trabalhadores em linha de produção. In: Proceedings of XL Simpósio Brasileiro de Pesquisa Operacional, João Pessoa, Brazil, pp. 143–152 (2008)
9. Kirkpatrick, S., Gellat, D.C., Vecchi, M.P.: Optimization by simulated annealing. Science 220, 671–680 (1983)
10. Baybars, I.: A survey of exact algorithms for the simple assembly line balancing problem. Management Science 32, 909–932 (1986)
11. Pinnoi, A., Wilhelm, W.E.: A family of hierarchical models for assembly system design. International Journal of Production Research 35, 253–280 (1997)
12. Miralles, C., García-Sabater, J.P., Andrés, C.: Application of u-lines principles to the assembly line worker assignment and balancing problem: a model and a solving procedure. In: Proceedings of the Operational Research Peripatetic Postgraduate Programme International Conference (ORP3), Valencia, Spain (2005)
13. Costa, A.M., Miralles, C.: Job rotation in assembly lines employing disabled workers. International Journal of Production Economics (forthcoming)
14. Glover, F.: Tabu search and adaptive memory programing. In: Barr, R.S., Helgason, R.V., Kennington, J.L. (eds.) Interfaces in Computer Science and Operations Research, pp. 1–75. Kluwer, Dordrecht (1996)
15. Mladenovic, N., Hansen, P.: Variable neighborhood search. Computers and Operations Research 24, 1097–1100 (1997)
16. Kilbridge, M., Wester, L.: The balance delay problem. Management Science 8(1), 69–84 (1961)
17. Scholl, A.: Data of assembly line balancing problems. University Press 16, TU Darmstadt (1993)
18. ILOG France: Ilog Cplex 10.0: user's manual (2006)
19. Chaves, A.A., Lorena, L.A.N.: Hybrid metaheuristic for the prize collecting travelling salesman problem. In: van Hemert, J., Cotta, C. (eds.) EvoCOP 2008. LNCS, vol. 4972, pp. 123–134. Springer, Heidelberg (2007)
20. Filho, G.R., Nagano, M.S., Lorena, L.A.N.: Evolutionary clustering search for flowtime minimization in permutation flow shop. In: Bartz-Beielstein, T., Blesa Aguilera, M.J., Blum, C., Naujoks, B., Roli, A., Rudolph, G., Sampels, M. (eds.) HCI/ICCV 2007. LNCS, vol. 4771, pp. 69–81. Springer, Heidelberg (2007)

An ELSxPath Relinking Hybrid for the Periodic Location-Routing Problem

Caroline Prodhon

Institut Charles Delaunay, Université de Technologie de Troyes, BP 2060,
10010 Troyes Cedex, France
caroline.prodhon@utt.fr

Abstract. The well-known Vehicle Routing Problem (VRP) has been deeply studied over the last decades. Nowadays, generalizations of VRP are developed toward tactical or strategic decision levels of companies. The tactical extension plans a set of trips over a multiperiod horizon, subject to frequency constraints. The related problem is called the Periodic VRP (PVRP). On the other hand, the strategic extension is motivated by interdependent depot location and routing decisions in most distribution systems. Low-quality solutions are obtained if depots are located first, regardless the future routes. In the Location-Routing Problem (LRP), location and routing decisions are simultaneously tackled. The goal here is to combine the PVRP and LRP into an even more realistic problem covering all decision levels: the Periodic LRP or PLRP. A hybrid evolutionary algorithm is proposed to solve large size instances of the PLRP. First, an individual representing an assignment of customers to combinations of visit days is randomly generated. Then, a heuristic based on the Randomized Extended Clarke and Wright Algorithm (RECWA) creates feasible solutions. The evolution operates through an Evolutionary Local Search (ELS) on visit days assignments. The algorithm is hybridized with a Path Relinking between individuals from an elite list. The method is evaluated on three sets of instances and solutions are compared to the literature on particular cases such as one-day horizon (LRP) or one depot (PVRP). This metaheuristic outperforms the previous methods for the PLRP.

Keywords: Heuristic, ELS, Path Relinking, Periodic Location-Routing Problem.

1   Introduction

Logistic systems involve different decision levels, usually tackled separately to reduce the complexity of the global problem. Nevertheless, Salhi and Rand [25] have shown that the strategy to solve the depot location then the vehicle routing, often leads to suboptimization. The Location-Routing Problem (LRP) appeared relatively recently in literature is a combination of both decisions. As shown in a survey [14], most early published papers consider either capacitated routes or capacitated depots [28,2], but not both except very recently [3,18,19,20,30].

M.J. Blesa et al. (Eds.): HM 2009, LNCS 5818, pp. 15–29, 2009.

Beside the strategic aspect of depot location, a focus on tactical decision on Vehicle Routing Problems (VRP) leads to consider some extensions as the periodic aspect of the customers' demands. The resulting problem is known as Periodic VRP or PVRP, introduced in 1984 by Christofides and Beasley [6]. It consists in integrating frequency constraints on visited customers over a given multiperiod horizon. Exact methods are available [11,15] and report optimal solutions on instances involving until 80 customers (asymmetric problem). However, the methods used to solve PVRP are mainly heuristics [5,6,27]. Powerful approaches are the tabu search algorithm proposed by Cordeau *et al.* [8] and a variable neighborhood search heuristic from Hemmelmayr *et al.* [10].

A combination of the LRP and the PVRP have been introduced for the first time in [22], resulting in an even more realistic problem: the periodic LRP or PLRP. The objective is to determine the set of depots to open, the combination of service days to assign to customers and the routes originating from each depot for each period of the horizon, in order to minimize the total cost. The first proposed method to solve it [22] is an iterative algorithm. In [23] and [24] genetic based metaheuristics are developed. In the first one [23], each iteration of the algorithm begins by assigning a fixed combination of service days to each customer, with respect to their required service frequency, for the entire set of individuals (solutions) from the population. Then, within this global iteration, the evolution handles the location and routing aspects by a Memetic Algorithm with Population Management (MA|PM) scheme. For each child, a local search occurs on the periodic decisions. This allows to record a possible better assignment of service days that would be used in the next global iteration of the method. In the second genetic based metaheuristic [24], the evolutionary algorithm manages the multi-period aspect by dealing with the assignment of the customers to service days. The fitness of an individual (total cost of the associated PLRP solution) is evaluated by a heuristic based on the Randomized Extended Clarke and Wright Algorithm (RECWA) [18]. Local searches try to improve the routing within a day or by finding a better combination of visit days to customers.

Based on the observation of the results from previous methods, it seems more promising to develop an algorithm deepening the search on the periodic aspect. In this paper, the proposed approach is a hybrid evolutionary algorithm. It combines an Evolutionary Local Search (ELS) [17,29] with a Path Relinking (PR) [9] to manages the assignment of the customers to service days. Then, it attributes the fitness $F(S)$ to individual S by evaluating the total cost of an associated PLRP feasible solution thanks to a heuristic based on the Randomized Extended Clarke and Wright Algorithm (RECWA) [19].

The paper is organized as follows. The problem is defined in more details in Section 2. Section 3 describes the framework of the proposed algorithm. Section 4 explains how to evaluate a solution. The performances of the method are evaluated in Section 5. Some concluding remarks close the paper.

2   Problem Definition

The PLRP studied in this paper is defined on an horizon H and a complete, weighted and undirected network $G = (V, E, C)$. V is a set of nodes comprised of a subset I of m possible depot locations and a subset $J = V \backslash I$ of n customers. c_{ij} is the traveling cost between any two nodes i and j. Each customer $j \in J$ has to be served with a given frequency over H, and $Comb_j$ is its set of possible combinaisons of serviced days. d_{jlr} is the demand of customer j on day l of combinaison $r \in Comb_j$. A capacity W_i and an opening cost O_i are associated with each depot site $i \in I$. A set K of identical vehicles of capacity Q is available each day. A vehicle used at least once from a depot during H incurs a fixed cost F and can perform one single route per day.

The following constraints must hold: i) each customer j must be served exclusively on each day l of exactly one of the combinaison $r \in Comb_j$ by one vehicle and at the amount d_{jlr}; ii) the number of routes performed by day must not exceed N; iii) each route must begin and end at the same depot within the same day and its total load must not exceed Q; iv) the total load of the routes assigned to a depot on any day $l \in H$ must fit the capacity of that depot.

The objective is to find which subset of depots to open, which combinaison of visit days to assign to each customers and which routes to perform, in order to minimize the total cost.

The PLRP is NP-hard since it reduces to the VRP when $m = 1$ and $|H| = 1$. It is much more combinatorial than the VRP and therefore, due to the size of the targeted instances, a metaheuristic is proposed. It is an evolutionary approach in which an individual is tackled by an ELS scheme hybridized with a PR. The next section describes the algorithm more in details.

3   ELS×PR for the PLRP

3.1   Evolutionary Algorithms

Evolutionary Strategies such as genetic algorithms have been successfully applied to vehicle routing problems, especially when hybridized with local search (memetic algorithms - MA) [12,16] and with a Population Management (MA|PM) [26]. Very good solutions have also been observed on applications to various extensions: production-distribution problem [4] or LRP [18] for instance.

Based on these encouraging results, MA|PM is applied on the PLRP in [23] and outperforms the previous algorithm dedicated to this problem [22]. The idea is to manage the routing part with the evolutionary tools, but the assignment of customers to combinations of service days is fixed for the entire set of individuals (solutions) from the population. Only when the algorithm refreshes the population (new global iteration of the method), a possible better assignment is used, based on the best one recorded during specific local search. This approach outperforms the algorithm from [22] by 1.93% on average on PLRP instances. However, another genetic scheme is proposed in [24] in which the serviced days are encoded within the chromosomes. This last method produces even better results by reducing of almost 5% the costs obtained with the algorithm from [22] on the PLRP instances.

In this paper, the aim is to develop an evolutionary algorithm focusing on the periodic aspect and searching more intensively each encountered attractive bassins. The metaheuristic applied to achieve this goal is a hybridization of an ELS with a PR.

3.2   Evolutionary Local Search

The choice of an ELS based heuristic is done to tackle the serviced combinaison assignments in a fast and efficient algorithm. ELS is an extension of the Iterated Local Search (ILS) [13]. The latter builds successive solutions, creating at each iteration one child-solution, thanks to mutation and local search. ELS is similar but generates $MaxMut > 1$ children-solutions at each iteration and selects the best one. The purpose of ELS is to better sample the attraction basins close to the current local optimum, before leaving it.

For the PLRP, an individual S is encoded as a vector of dimension n and the information recorded at index i corresponds to the combinaison of service days assigned to customer i.

The fitness $F(S)$ of individual S is the total cost of an associated PLRP solution. To evaluate $F(S)$, a heuristic "$Eval$" (see section 4), based on the Randomized Extended Clarke and Wright algorithm (RECWA), is applied [19].

The evolution is made by an ELS. At each global iteration, a function called "$Diversification$" generates an initial individual S. The principle is to modify the best current solution with random exchanges of the visit days assignment on a percentage $Mut1$ of customers.

Then, an inner loop searches the solution space in the neighborhood of S. First, S undergoes $MaxMut$ independent mutations. A mutation here is a random modification as in $Diversification$ but for a percentage $Mut2 < Mut1$ of customers. This leads to individual S'' evaluated by $Eval$. A local search LS2 (see section 4.3) intends to reduce the routing cost by finding a better combination of visit days for customers. Second, the best solution S' among the initial one and the $MaxMut$ generated individuals is kept as the new start point for the next ELS iteration.

After $MaxELSIt$ iterations of the inner loop, an intensification occurs thanks to a PR between the best solution of the global iteration and a solution from an elite list. More details about PR are given in 3.3. Then, the next global iteration begins by a new call to the $Diversification$ function generating a new individual. The algorithm stops when a given number $MaxTotIt$ of solutions resulting of mutation has been evaluated.

3.3   Path Relinking

A PR can be applied to most metaheuristics as a kind of intensification or post-optimization: new solutions are generated by exploring trajectories that connect a small subset of high-quality solutions collected during the search.

Each trajectory is obtained by introducing in a solution S some attributes from another solution T called the guiding solution [9].

In this paper, PR is realized as an intensification, between each locally optimal solution S from each global iteration of the ELS×PR, and another one T. The latter is chosen among a group of elite solutions called *Elite*, composed of best solutions found at the end of each ELS×PR global iteration. More precisely, the selected individual T is the more distant to S. Let $D(S,T)$ be a measure of the differences between S and T, and $S(i)$ and $T(i)$ be the assignment to combinaison of visit days for customer i in respectively S and T. For each $i \in J$, when $S(i) \neq T(i)$, count 1 in the distance.

Thus, from the couple (S,T), S is transformed step by step into T. The attributes introduced in S are the visit days assignments, and they are taken in the order encountered in T (from customer 1 to n).

Each solution obtained after the introduction of β attributes undergoes an evaluation through *Eval* and LS2.

Figure 1 gives an overview of the proposed method.

4 The Evaluation of an Individual

To evaluate the cost of the generated individuals, a function *Eval* based on an algorithm (RECWA) dedicated to the location-routing problem is implemented.

4.1 The Randomized Extended Clarke and Wright Algorithm

The Clarke and Wright algorithm (CWA in the sequel) is a saving algorithm originally designed for the VRP [7].

The extended version (ECWA) for the LRP computes an initial solution by assigning each customer to their closest depot in which they can fit. As in the original CWA, a dedicated route is created to link the customer to its selected depot. When all customers are assigned, the depots without customers are closed, giving an initial solution looking like a bunch of flowers.

When two routes R and S are merged in ECWA, the resulting route T may be reassigned to the depot r of R, to the depot s of S, or to another depot t, opened or not (lower part of Figure 2). Hence, $4m$ mergers must be evaluated for each pair of routes.

The randomized version of ECWA is used here. The idea is to create a restricted candidate list (RCL) of size α from the savings calculated during the merger evaluations, and randomly choose one among this list. The elements of the RCL are the pairs of routes giving the α best savings.

RECWA applied as explained above is said in its *diversification mode*. The aim is to explore the solution space by allowing the entire set of depots I to be used.

The *intensification mode* restricts the level of randomness : only a subset SD of depots is allowed to be opened in RECWA during the construction of the bunch of flowers and the mergers.

```
 1: BestCost := +∞      NbIt := 0
 2: Create a random individual S
 3: repeat
 4:     Diversification(S)
 5:     Eval(S)
 6:     LS2(S)
 7:     if F(S) < BestCost then
 8:         BestCost := F(S)
 9:         BestSoln := S
10:     end if
11:     BestCostMut := F(S)
12:     NbELS := 0
13:     repeat
14:         NbELS := NbELS + 1
15:         S' := S
16:         NbMut := 0
17:         repeat
18:             NbIt := NbIt + 1      NbMut := NbMut + 1
19:             S'' := S'
20:             Mutation(S'')
21:             Eval(S'')
22:             LS2(S'')
23:             if F(S'') < BestCostMut then
24:                 BestCostMut := F(S'')
25:                 S := S''
26:             end if
27:             if F(S'') < BestCost then
28:                 BestCost := F(S'')
29:                 BestSoln := S''
30:             end if
31:         until NbMut ≥ MaxMut AND NbIt ≥ MaxTotIt
32:     until NbELS ≥ MaxELSIt AND NbIt ≥ MaxTotIt
33:     Add S in Elite
34:     Dist := 0
35:     for all T' ∈ Elite do
36:         if D(S, T') > Dist then
37:             Dist := D(S, T')
38:             T := T'
39:         end if
40:     end for
41:     if Dist > 0 then
42:         PR(S,T)
43:     end if
44: until NbIt ≥ MaxTotIt
45: Return (BestSoln)
```

Fig. 1. Algorithm of the hybrid ELS×PR for the Periodic Location-Routing Problem

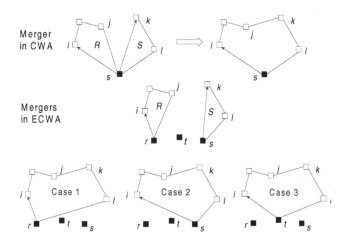

Fig. 2. Mergers in CWA and ECWA

At the end of the RECWA, a feasible solution is available. It is improved by a local search LS dedicated to the LRP from [19], applied on each day in H. Note that a depot can only be opened by RECWA, not by this local search, and that the assignment to service day is not managed here.

More details about RECWA can be found in [19].

4.2   Depot Location

RECWA is an algorithm dedicated to LRP that deals with the routing and the location decisions. It is used to evaluate each individual from ELS. However, it is not possible to simply apply RECWA on each day of the horizon. Indeed, such a strategy would open different sets of depots on each period of H. This may not be the best approach since the set up cost induced would highly increase.

Thus, a first run on each day is done independently to provide information about the well suited depots according to the visit days assignment of the customers given by the individual. For each depot d, a ratio $R1_d = \frac{MaxC_d}{W_d}$ is calculated, where $MaxC_d$ is the maximal sum of the demand assigned to d on one day over H, as well as the ratio $R2_d = \frac{Dem_d}{DemTot}$, where Dem_d is the sum of the demand assigned to d over H and $DemTot$ is the total demand of the n customers. The higher the value of $R_d = R1_d \times R2_d$, the more suited the depot d over H.

To avoid the opening of too many depots, a subset SD of available depots is defined. SD is made of depots picked in I on the base of the above information. Then, the routing is done only from SD during a new call to RECWA running here in its intensification mode. If a customer cannot be served by lack of capacity in the depots from SD, it can add its closest depot in the set.

To decide which depots enter in SD, R_d is compared to R ($R = \frac{\sum_{d=1..m} R_d}{m}$). First let $SD = \emptyset$. For each $d \in I$, if $R_d \geq R$, depot d is added to SD with probability $prob$. If the $SumCapaSD$ (the sum of the capacities of the depots in SD) is not greater to $MaxDemDay$ (the maximal demand over H of the set of customers assigned to each period), additional depots, randomly selected, are added to the subset until $SumCapaSD \geq MaxDemDay$.

Figure 3 gives an overview of evaluation of an individual.

```
 1: S an individual
 2: SD := I
 3: for each day ∈ H do
 4:     RECWA(S,day,SD)
 5: end for
 6: R :=0
 7: SD := ∅
 8: for each d ∈ I do
 9:     R_d := R1_d × R2_d
10:     R := R + R_d
11: end for
12: R := R/m
13: SumCapaSD := 0
14: for each d ∈ I do
15:     if R < R_d AND rand()< prob then
16:         SD := SD ∪ {d}
17:         SumCapaSD := SumCapaSD + W_d
18:     end if
19: end for
20: if SumCapaSD < MaxDemDay then
21:     repeat
22:         Randomly choose a depot d not in SD
23:         SD := SD ∪ {d}
24:         SumCapaSD := SumCapaSD + W_d
25:     until SumCapaSD ≥ MaxDemDay
26: end if
27: for each day ∈ H do
28:     RECWA(S,day,SD)
29: end for
30: Return (F(S))
```

Fig. 3. Algorithm of the evaluation of an individual from the ELS

4.3   Local Search for the PLRP

The algorithm works on the evolution of high-quality individuals. This quality results from the application of a specific PLRP local search procedure LS2 on the solutions deduced from the individual.

It intends to reduce the routing cost of a child by finding a new combination of visit days to customers. A move is performed if the best insertion cost of a customer in the days of its new combination is lower than the cost to serve it at its current position. Of course, the capacity constraints must hold.

5   Computational Study

5.1   Instances

Our ELS×PR hybrid is evaluated on three sets of randomly generated Euclidean instances. The two first ones can be downloaded at [21]. They are made of 30 instances with a set of homogenous capacitated vehicles and a set of possible capacitated depots. Their main characteristics are the followings: number of depots $m \in \{5, 10\}$, number of clients $n \in \{20, 50, 100, 200\}$ vehicle capacity $Q \in \{70, 150\}$ and number of clusters $\beta \in \{0, 2, 3\}$. The case $\beta = 0$ corresponds in fact to a uniform distribution of customers in the Euclidean plane. These instances in which all numbers are integer were generated as follows. For given choices of m, n, Q, and β, the customers' locations are randomly chosen in the Euclidean plane, and the traveling costs c_{ij} correspond to the distances, multiplied by 100 and rounded up to the nearest integer. Each demand follows a uniform distribution in interval $[10, 20]$.

Table 1. Allowed set of combinations of visit days

Frequency	Combinations of visit days
1	Monday
	Tuesday
	Wednesday
	Thursday
	Friday
2	Monday - Wednesday
	Monday - Thursday
	Tuesday - Friday
3	Monday - Wednesday - Friday

To mimic a working week, *the first set*, especially generated for the PLRP, is made of a 7-day cyclic horizon H with 2 idle days (Saturday and Sunday). The visit frequency of the customers $s(i)$ is between once and three times a week. The allowed set of combinations of visit days $comb(i)$ are given in Table 1 and forbids visits on two consecutive days. The demand d_{jlr} of customer j on each day l depends on the chosen service combination r. It varies with the number of days separating two visits, but it can be pre-calculated. The demand over the horizon is divided by the number of service days, leading to an average demand by day. Then, for each service day l of a combination r, d_{jlr} is this average demand multiplied by the number of days since the last service day. For the first day of the combination, the d_{jlr} is the difference of the demand over the horizon minus the demand served on each other day from the combination ($\sum_{k\neq l} d_{jkr}$). The results obtained on these instances are compared with respect to the results proposed in [23,24].

The second set contains LRP instances, created for our previous works [19,20]. It is reused in this study to compare the performances of proposed method with respect to the best-known results (BKR), obtained when testing various methods with different parameters and genetic algorithms from [23,24].

Finally, *the third set* comprises 30 instances for the PVRP available at [1]. The original set has 32 instances, but 2 are discarded because they contain customers having the same visit frequency but not the same combination of visit days. The horizon is made of P periods and the demand is equal in each service day. The fleet size is limited to k and each vehicle has a capacity Q. The number of customers n ranges from 20 to 417. The traveling costs are equal to the Euclidean distances (not rounded). The performances of our algorithm are compared with the best-known results (BKR) provided at [1] on these instances and results from [23,24].

5.2   Implementation, Parameters and Algorithms Compared

The proposed algorithm is coded in Visual C++ and has been tested on a Dell Latitude D420, with an Intel Centrino Duo 1.2 GHz, 1 GB of RAM and running Windows XP.

The following parameters have been selected after a testing phase, in order to provide the best average solution values:

$MaxMut = 6$, $Mut1 = 0.35$, $Mut2 = 0.05$, $MaxELSIt = 5$, $MaxTotIt = 30 \times (1 + P)$, $\alpha = 7$, $\beta = n \times 0.02$ and $prob = 0.85$.

5.3   Discussion of Results

In the following tables, each line corresponds to an instance or a subset of instances having the same characteristics. Column ELS×PR indicate the average value of the objective function over 5 runs and on the subset and times CPU are given in seconds. The columns reported as Gap_{xxx} are the deviation in percentage between the proposed result and the xxx solution. BKS are the Best-Known Solutions. On PRLP, they are the best solution found by previous methods ([22,23,24]). On LRP and PVRP instances, they come from the respective websites [21,1]. $MAPM$ refers to the method from [23] and Gen to the genetic algorithm from [24].

First Set. Table 2 provides a comparison between the proposed ELS×PR and our earlier evolutionary metaheuristics detailed in [23] and [24] on the 30 PLRP instances from the first set. In $MAPM$, each global iteration of the algorithm begins by allocating each customer to a combination of service days while in Gen, the genetic scheme searches the most promising assignment.

Table 2. Results on the Periodic Location-Routing Problem

Instance	$ELS \times PR$	CPU	Gap_{BSK}	Gap_{MAPM}	Gap_{Gen}
P20.5.0a	**80546.3**	1.1	-1.97	-7.65	-1.97
P20.5.0b	78564.7	1.3	1.39	1.39	-4.51
P20.5.2a	**81827.6**	1.1	-0.31	-1.17	-0.31
P20.5.2b	**63492.1**	1.1	-2.42	-8.22	-2.42
P50.5.0a	**158675.6**	5.5	-1.96	-5.32	-1.96
P50.5.0b	**144072.4**	9.1	-5.01	-8.20	-5.01
P50.5.2a	153138.6	6.6	1.68	-5.09	1.68
P50.5.2b	122152.8	8.3	1.43	1.43	-2.20
P50.5.2a'	**180679.9**	7.5	-4.58	-6.22	-4.58
P50.5.2b'	**103866.5**	7.6	-8.63	-14.47	-8.63
P50.5.3a	**163986.4**	7.4	-4.78	-9.60	-4.78
P50.5.3b	**116288.4**	7.7	-1.85	-5.91	-1.85
P100.5.0a	**371290.7**	40.0	-0.67	-2.86	-0.67
P100.5.0b	**239804.2**	41.2	-3.19	-3.19	-3.92
P100.5.2a	292208.9	49.1	1.78	-4.56	-1.18
P100.5.2b	**177294.5**	41.4	-0.87	-9.97	-0.87
P100.5.3a	**240650.0**	32.4	-0.96	-7.67	-0.96
P100.5.3b	**182130.2**	41.9	-2.92	-13.73	-2.92
P100.10.0a	**284559.8**	56.0	-2.80	-9.15	-2.80
P100.10.0b	**227180.9**	78.9	-2.67	-4.21	-2.67
P100.10.2a	**286286.7**	70.1	-0.67	-8.30	-2.52
P100.10.2b	**181069.6**	84.7	-4.47	-10.46	-4.47
P100.10.3a	**277844.9**	72.4	-3.19	-8.46	-3.02
P100.10.3b	**211735.2**	92.3	-2.58	-6.36	-0.29
P200.10.0a	**480716.3**	351.1	-1.22	-14.03	-2.69
P200.10.0b	406137	547.3	0.43	-11.49	-0.50
P200.10.2a	**414964.3**	298.5	-1.44	-3.99	-3.35
P200.10.2b	334163.3	644.3	3.10	-6.55	-2.17
P200.10.3a	**575764.0**	426.7	-2.11	-7.45	-4.08
P200.10.3b	366447.6	460.1	2.76	-3.92	-3.79
Average		116.4	-1.62	-6.85	-2.65

The results show that the proposed approach performs better than both compared methods while the CPU time is divided by almost $1/3$. Furthermore, it reaches a gap of 1.62% with BKS. More precisely, it is able to improve 23 over 30 of the already best-known solutions.

In comparison with $MAPM$, an improvement of 6.85% of the cost on average is observed. The very good performances attest the importance on the choice of assignment to combinaison of visit days. The proposed ELS×PR seems to hardly manage that point since it even outperforms Gen by 2.65% while the latter also manages the periodic aspect with an evolutionary technique.

Second Set. Table 3 gives the average results on the 30 LRP instances from the second set. The gap with previous methods is positive but remains small. This is due to the fact that the ELS and even the PR focus on the periodic aspect and $|H| = 1$ on these instances. Thus, only the evaluation based on RECWA contributes to the performance of the method on location and routing decisions. However, the proposed metaheuristic is able to provide good solutions even on other problems than the one it is dedicated to (gap with BKS is 4.82% on average) and some best-known solutions are reached on small instances. Furthermore, it can obtain better solution than $MAPM$ and outperforms Gen with an average gap at -0.35% while the CPU time is almost divided by a third.

Table 3. Results on the Location-Routing Problem

n	m	$ELS \times PR$	CPU	Gap_{BKS}	Gap_{MAPM}	Gap_{Gen}
20	5	45124.5	0.4	0.07	-1.05	0.03
50	5	76664.9	5.2	3.58	1.55	0.76
100	5	202352.8	39.3	1.57	-1.61	-1.45
100	10	260618.8	71.5	9.18	1.80	-1.15
200	10	455915.1	546.7	8.54	5.28	-0.17
Average			132.94	4.82	1.37	-0.35

Table 4. Results on the Periodic Vehicle Routing Problem

I	n	k	P	$ELS \times PR$	CPU	Gap_{BKS}	Gap_{MAPM}	Gap_{Gen}
1-6	50-75	1-6	2-10	1065.4	10.7	5.1	-2.8	-2.2
7-13	100-417	2-8	1-9	1843.8	59.1	7.9	1.4	-2.9
14-22	20-184	4	2-6	3382.5	20.3	1.4	-0.9	-0.9
24-32	51-153	6	3-9	38275.0	22.0	7.5	-2.4	-1.8
Average					26.6	5.26	-1.27	-1.82

Third Set. Table 4 provides a comparison on the PVRP instances from the third set, between the results obtained by the proposed ELS×PR and the best-known solutions (BKR) available from [1] and previous PLRP methods. Note that our algorithm does not manage the fleet size. Thus the obtained solution may not succeed in finding a solution compatible with the limited number of vehicles (this does not happened frequently).

Gap with BKS is +5.26% on average showing how the periodic aspect is hard to deal with. However, the proposed results clearly outperform $MAPM$ with a negative gap at 1.27% and Gen with an improvement of 1.82% while requiring about a half of CPU time. This confirms the conclusion made on PLRP instances on the importance on the choice of assignment to combinaison of visit days.

5.4   Further Analysis

It may be interesting to have a look at the stability of the method over the 5 runs. For each PLRP instance, the maximal, minimal and average obtained values are normalized with respect to the average cost. Thus, we obtain the disparity shown on Figure 4.

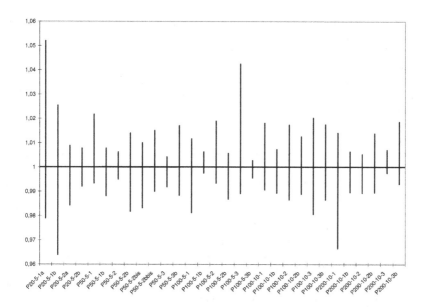

Fig. 4. Stability of the method over the 5 runs

We can see that the randomness does not have a great impact on the stability of the results since only 1% of variation is observed on average.

6 Conclusion

In this paper, a new metaheuristic for the Periodic Location Routing Problem (PLRP) with both capacitated depots and vehicles is presented. The method is an hybrid evolutionary algorithm called ELS×PR. It is an Evolutionary Local Search method hybridized with a Path Relinking which manage the period aspect. The method has been tested on three sets of small, medium and large scale instances, and compared to other heuristics on various kind of instances such as one-day horizons (LRP) or one depot (PVRP). The solutions obtained show that this algorithm is able to find good solutions and clearly outperforms the previous methods dedicated to the PLRP.

References

1. http://neo.lcc.uma.es/radi-aeb/WebVRP/ (2007)
2. Albareda-Sambola, M., Díaz, J.A., Fernández, E.: A compact model and tight bounds for a combined location-routing problem. Computers and Operations Research 32, 407–428 (2005)
3. Belenguer, J.M., Benavent, E., Prins, C., Prodhon, C., Wolfler-Calvo, R.: A cutting plane method for the capacitated location-routing problem. In: Odysseus 2006 - Third International Workshop on Freight Transportation and Logistics, Altea, Spain, May 2006, pp. 50–52 (2006)

4. Boudia, M., Louly, M.A.O., Prins, C.: A memetic algorithm with population management for a production-distribution problem. In: Dolgui, A., Morel, G., Pereira, C.E. (eds.) Preprints of the 12th IFAC Symposium on Information Control Problems in Manufacturing INCOM 2006, Saint Etienne-France, May 2006, vol. 3, pp. 541–546 (2006)
5. Chao, M., Golden, B.L., Wasil, E.: An improved heuristic for the periodic vehicle routing problem. Networks 26, 25–44 (1995)
6. Christofides, N., Beasley, J.E.: The period routing problem. Networks 14, 237–256 (1984)
7. Clarke, G., Wright, J.W.: Scheduling of vehicles from a central depot to a number of delivery points. Operations Research 12, 568–581 (1964)
8. Cordeau, J.F., Gendreau, M., Laporte, G.: A tabu search heuristic for periodic and multi-depot vehicle routing problems. Networks 30, 105–119 (1997)
9. Glover, F., Laguna, M., Martí, R.: Fundamentals of Scatter Search and Path Relinking. Control and Cybernetics 39, 653–684 (2000)
10. Hemmelmayr, V.C., Doerner, K.F., Hartl, R.F.: A variable neighborhood search heuristic for periodic routing problems. European Journal of Operational Research 195(3), 791–802 (2009)
11. Laporte, G., Norbert, Y., Arpin, D.: Optimal solutions to capacitated multi-depot vehicle routing problems. Congressus Numerantium 44, 283–292 (1984)
12. Lima, C.M.R.R., Goldbarg, M.C., Goldbarg, E.F.G.: A memetic algorithm for heterogeneous fleet vehicle routing problem. Electronic Notes in Discrete Mathematics 18, 171–176 (2004)
13. Lourenço, H.R., Martin, O., Stützle, T.: Iterated Local Search. In: Glover, F., Kochenberger, G. (eds.) Handbook of Metaheuristics, pp. 321–353. Kluwer, Boston (2003)
14. Min, H., Jayaraman, V., Srivastava, R.: Combined location-routing problems: a synthesis and future research directions. European Journal of Operational Research 108, 1–15 (1998)
15. Mingozzi, A., Valletta, A.: An exact algorithm for period and multi-depot vehicle routing problems. In: Odysseus 2003 - Second International Workshop on Freight Transportation and Logistics, Palermo, Italy (May 2003)
16. Prins, C.: A simple and effective evolutionary algorithm for the vehicle routing problem. Computers and Operations Research 31, 1985–2002 (2004)
17. Prins, C.: A GRASP × Evolutionary Local Search Hybrid for the Vehicle Routing Problem. In: Bio-inspired Algorithms for the Vehicle Routing Problem, Studies in Computational Intelligence, pp. 35–53. Springer, Berlin (2009)
18. Prins, C., Prodhon, C., Wolfler-Calvo, R.: A memetic algorithm with population management (MA|PM) for the capacitated location-routing problem. In: Gottlieb, J., Raidl, G.R. (eds.) EvoCOP 2006. LNCS, vol. 3906, pp. 183–194. Springer, Heidelberg (2006)
19. Prins, C., Prodhon, C., Wolfler-Calvo, R.: Solving the capacitated location-routing problem by a GRASP complemented by a learning process and a path relinking. 4OR - A Quarterly Journal of Operations Research 4, 221–238 (2006)
20. Prins, C., Prodhon, C., Wolfler-Calvo, R.: Solving the capacitated location-routing problem by a cooperative lagrangean relaxation-granular tabu search heuristic. Transportation Science 41(4), 470–483 (2007)
21. Prodhon, C.: http://prodhonc.free.fr/homepage (2007)
22. Prodhon, C.: An iterative metaheurtistic for the periodic location-routing problem. In: GOR 2007, Saarbrücken (September 2007)

23. Prodhon, C., Prins, C.: A memetic algorithm with population management (MA|PM) for the periodic location-routing problem. In: Blesa, M.J., Blum, C., Cotta, C., Fernández, A.J., Gallardo, J.E., Roli, A., Sampels, M. (eds.) HM 2008. LNCS, vol. 5296, pp. 43–57. Springer, Heidelberg (2008)
24. Prodhon, C.: An evolutionary algorithm for the periodic location-routing problem. In: Odysseus 2009 - Third International Workshop on Freight Transportation and Logistics, Cesme, Turkey, May 2009, pp. 43–57 (2009)
25. Salhi, S., Rand, G.K.: The effect of ignoring routes when locating depots. European Journal of Operational Research 39, 150–156 (1989)
26. Sörensen, K., Sevaux, M.: MA | PM: memetic algorithms with population management. Computers and Operations Research 33, 1214–1225 (2006)
27. Tan, C.C.R., Beasley, J.E.: A heuristic algorithm for the periodic vehicle routing problem. OMEGA International Journal of Management Science 12(5), 497–504 (1984)
28. Tuzun, D., Burke, L.I.: A two-phase tabu search approach to the location routing problem. European Journal of Operational Research 116, 87–99 (1999)
29. Wolf, S., Merz, P.: Evolutionary local search for the super-peer selection problem and the p-hub median problem. In: Bartz-Beielstein, T., Blesa Aguilera, M.J., Blum, C., Naujoks, B., Roli, A., Rudolph, G., Sampels, M. (eds.) HCI/ICCV 2007. LNCS, vol. 4771, pp. 1–15. Springer, Heidelberg (2007)
30. Wu, T.H., Low, C., Bai, J.W.: Heuristic solutions to multi-depot location-routing problems. Computers and Operations Research 29, 1393–1415 (2002)

Hybridizing Beam-ACO with Constraint Programming for Single Machine Job Scheduling

Dhananjay Thiruvady[1,3], Christian Blum[2], Bernd Meyer[1], and Andreas Ernst[3]

[1] Clayton School of Information Technology, Monash University, Australia
{dhananjay.thiruvady,bernd.meyer}@infotech.monash.edu.au
[2] ALBCOM Research Group, Universitat Politècnica de Catalunya, Barcelona, Spain
cblum@lsi.upc.edu
[3] CSIRO Mathematics and Information Sciences, Australia
andreas.ernst@csiro.au

Abstract. A recent line of research concerns the integration of ant colony optimization and constraint programming. Hereby, constraint programming is used for eliminating parts of the search tree during the solution construction of ant colony optimization. In the context of a single machine scheduling problem, for example, it has been shown that the integration of constraint programming can significantly improve the ability of ant colony optimization to find feasible solutions. One of the remaining problems, however, concerns the elevated computation time requirements of the hybrid algorithm, which are due to constraint propagation. In this work we propose a possible solution to this problem by integrating constraint programming with a specific version of ant colony optimization known as Beam-ACO. The idea is to reduce the time spent for constraint propagation by parallelizing the solution construction process as done in Beam-ACO. The results of the proposed algorithm show indeed that it is currently the best performing algorithm for the above mentioned single machine job scheduling problem.

1 Introduction

The integration of exact methods with metaheuristics based on the construction of solutions has emerged as a promising approach to combinatorial optimization [4,11], because most metaheuristics in their basic form are unable to deal with non-trivial hard constraints. Using naïve constraint handling techniques (such as penalty methods) requires a significant amount of fine-tuning that can only be done by trial-and-error and more often than not leads to severe performance problems [8]. More sophisticated constraint handling approaches, such as repair techniques or decoder-based approaches, have to be designed on a case-by-case basis. This is a costly and error-prone process. In contrast to this, a generic integration of exact constraint handling techniques with a metaheuristic allows to systematically customize the metaheuristic for specific problem domains through a well-defined declarative specification of the problem model [14].

M.J. Blesa et al. (Eds.): HM 2009, LNCS 5818, pp. 30–44, 2009.

Such algorithms use a constraint propagation method to systematically prune the search space and thus to increase the efficiency of the underlying search method.

In [14] we showed that the hybridization of ant colony optimization (ACO) [10] and constraint programming (CP) [13] was effective on a single machine job scheduling problem. However, this algorithm had run-time problems due to the overhead of constraint propagation and obtaining high quality solutions for large problem instances in a reasonable time-frame was not possible. The main problem here being that the ACO component of the algorithm frequently reconstructs (parts of) a solution requiring repetition of the same propagation steps. In this study we (partially) overcome this problem by replacing the standard ACO component with Beam-ACO [3]. This algorithm is itself a hybrid between ACO and Beam search [15,19], which is an incomplete tree search method that propagates a search front of open nodes in parallel through the tree. In general, Beam-ACO works as any other ACO algorithm. At each iteration candidate solutions are probabilistically constructed on the basis of a so-called pheromone model, which is a set of numerical values that encode the algorithms' search experience. However, instead of the independent construction of solutions, Beam-ACO employs the parallel and non-independent construction of solutions at each iteration in the style of beam search.

Our results for the single machine scheduling problem with sequence dependent setup times show that the newly proposed hybrid method significantly reduces the amount of computation time in comparison to the method proposed in [14]. Through the partial parallelization of the search it allows to store intermediate solver states compactly in a systematic form. This efficient use of the constraint solver leads to more computation time that can be devoted to the optimization/learning component.

The paper is organized as follows. In Sec. 2, the single machine scheduling problem with sequence-dependent setup times is introduced. Sec. 3 is devoted to outlining the basic ACO algorithm considered in this paper. While Sec. 4 deals with the extension of this basic ACO algorithm to Beam-ACO, Sec. 5 explains the hybridization of both ACO variants with constraint programming. Finally, in Sec. 6 we provide an experimental evaluation of the proposed algorithms and in Sec. 7 we offer conclusions and an outline of future work.

2   Single Machine Job Scheduling with Setup Times

The single machine job scheduling (SMJS) problem with sequence-dependent setup times has received considerable interest within the AI and OR communities. It is closely related to other problems such as, for example, traveling salesman problems with time windows. In fact, the problem is also referred to as the asymmetric traveling salesman problem with time windows [2]. Deciding if a feasible solution exists is known to be NP-Complete [16] and finding an optimal solution is NP-Hard [2].

The problem is formally defined as follows. A single machine must process n jobs $J = \{1, ..., n\}$ with processing times

$\overline{R} = \{\overline{r_1}, \ldots, \overline{r_n}\}$, and due times $\overline{D} = \{\overline{d_1}, \ldots, \overline{d_n}\}$.[1] The machine cannot process more than one job at the same time. Moreover, the processing of an job must not be interrupted. Furthermore, for each ordered pair (i, j) of jobs (where $i \neq j$) a setup time $\overline{st_{ij}}$ is given. Any permutation π of the n jobs represents a (not necessarily feasible) solution to the problem. Given such a permutation π, where π_i denotes the operation at position i, the so-called starting times $\hat{S} = \{\hat{s}_1, \ldots, \hat{s}_n\}$ and the ending times $\hat{E} = \{\hat{e}_1, \ldots, \hat{e}_n\}$ are well-defined. They can be determined recursively in the following way: $\hat{s}_{\pi_1} = 0$ and $\hat{s}_{\pi_i} = \max\{\hat{e}_{\pi_{i-1}} + \overline{st_{\pi_{i-1}\pi_i}}, \overline{r_{\pi_i}}\}$, where $\hat{e}_{\pi_{i-1}} = \hat{s}_{\pi_{i-1}} + \overline{p_{\pi_{i-1}}}$. The objective is to find a permutation π^* such that $f(\pi^*) = \hat{e}_{\pi_n^*}$ is minimal. Function $f()$ is commonly called the *makespan*.

3 $\mathcal{MAX}$–$\mathcal{MIN}$ Ant System

Ant colony optimization (ACO) is a reinforcement based stochastic metaheuristic inspired by the foraging behaviour of real ants [10]. Foraging ants that return to the nest after finding a food source mark their return paths with pheromones. Other colony members probabilistically follow these pheromone trails. As pheromone deposits can be modulated by food or path quality, better paths attract more ants, which in turn deposit more pheromone on these paths. This positive feedback loop allows the colony to converge on paths leading to better food sources [6]. ACO algorithms use these principles to solve combinatorial optimization problems. Conceptually, each "ant" constructs a candidate solution in each iteration of the algorithm. At each step it extends its partial candidate solution by adding a new solution component until a complete solution is obtained. The next component is chosen probabilistically according to the pheromone information. In the context of the SMJS problem, solution components are jobs, a complete solution is a permutation/sequence of all jobs, and the virtual pheromones are used to learn job successor relations that are frequently found in high quality schedules. More specifically, the pheromone model $\mathcal{T}$ consists of a pheromone value τ_{ij} for each ordered pair (i, j) of jobs, where $i \neq j$. Additionally, pheromone model $\mathcal{T}$ consists of pheromone values τ_{0i}, $i = 1, \ldots, n$, where τ_{0i} represents the goodness of placing job i at the first position of the permutation under construction.

In this study we used the $\mathcal{MAX}$–$\mathcal{MIN}$ Ant System ($\mathcal{MMAS}$) [17] implemented in the hypercube framework [5]. The reason for choosing $\mathcal{MMAS}$, in contrast to the choice of ACS in [14], is that $\mathcal{MMAS}$ seems better suited than ACS for the hybridization with beam search. In the following we give, for space reasons, only a short technical description of this algorithm. For a more detailed introduction to the parameters and the functioning of this algorithm we refer to [5]. At each iteration, $n_a = 10$ artificial ants construct (not necessarily feasible) solutions to the problem in form of permutations of all jobs. The details of the algorithmic framework shown in Alg. 1 are explained in the following.

---

[1] Bars indicate constants.

Algorithm 1. $\mathcal{MMAS}$ for the SMJS problem

1: INPUT: A SMJS instance
2: $\pi^{bs} :=$ NULL, $\pi^{rb} :=$ NULL, $cf := 0$, $bs\_update :=$ FALSE
3: **forall** $\tau_{ij} \in \mathcal{T}$ **do** $\tau_{ij} := 0.5$ **end forall**
4: **while** termination conditions not satisfied **do**
5:     $S_{iter} := \emptyset$
6:     **for** $j = 1$ to n_a **do**
7:         $\pi_j :=$ ConstructPermutation()
8:         **if** π_j is a feasible solution **then** $S_{iter} := S_{iter} \cup \{\pi_j\}$ **endif**
9:     **end for**
10:    $\pi^{ib} :=$ argmin$\{f(\pi)|\pi \in S_{iter}\}$
11:    **if** π^{ib} is a feasible solution **then**
12:        Update($\pi^{ib}, \pi^{rb}, \pi^{bs}$)
13:        ApplyPheromoneValueUpdate($cf, bs\_update, \mathcal{T}, \pi^{ib}, \pi^{rb}, \pi^{bs}$)
14:        $cf :=$ ComputeConvergenceFactor($\mathcal{T}$)
15:        **if** $cf \geq 0.99$ **then**
16:            **if** $bs\_update =$ TRUE **then**
17:                **forall** $\tau_{ij} \in \mathcal{T}$ **do** $\tau_{ij} := 0.5$ **end forall**
18:                $\pi^{rb} :=$ NULL, $bs\_update :=$ FALSE
19:            **else**
20:                $bs\_update :=$ TRUE
21:            **end if**
22:        **end if**
23:    **end if**
24: **end while**
25: OUTPUT: π^{bs}

ConstructPermutation(): A permutation π of all jobs is incrementally constructed from left to right. The solution construction stops when either the permutation is complete, or when it becomes clear that the resulting solution will be infeasible. Given a partial permutation π with $i - 1$ positions already filled, a job $k \in J \setminus \{\pi_1, \ldots, \pi_{i-1}\}$ for the i-th position in π is chosen as follows:

$$\mathbf{p}(\pi_i = k) = \frac{\tau_{\pi_{i-1}k} \times \eta_k}{\sum_{j \in J \setminus \{\pi_1, \ldots, \pi_{i-1}\}} \left(\tau_{\pi_{i-1}j} \times \eta_j\right)} , \tag{1}$$

where η_k is a measure of the heuristic goodness for placing job k on position i of permutation π. Here η_k is defined as the inverse of the due time of job k, i.e., $1/\overline{d_k}$, until a feasible solution is found. Once this happens η_k, respectively η_j, is exchanged by $\eta_{\pi_{i-1}k}$, respectively $\eta_{\pi_{i-1}j}$, and is defined as the inverse of the setup time between jobs π_{i-1} and k, respectively j. The first construction step, that is, when π is still empty, is a special one. In this case, the probability $\mathbf{p}(\pi_1 = k)$ is proportional to the pheromone value τ_{0k}. When a feasible solution is found there is no heuristic bias applied in the first selection step.

Update($\pi^{ib}, \pi^{rb}, \pi^{bs}$): This procedure (as well as the pheromone update, see below) is only executed in case π^{ib} is a valid solution. It sets π^{rb} and π^{bs} to π^{ib} (i.e., the iteration-best solution), if $f(\pi^{ib}) < f(\pi^{rb})$ and $f(\pi^{ib}) < f(\pi^{bs})$.

Table 1. The schedule used for values κ_{ib}, κ_{rb} and κ_{bs} depending on cf (the convergence factor) and the Boolean control variable bs_update

| | $bs_update = $ FALSE | | | | $bs_update = $ TRUE |
	$cf < 0.4$	$cf \in [0.4, 0.6)$	$cf \in [0.6, 0.8)$	$cf \geq 0.8$	
κ_{ib}	1	2/3	1/3	0	0
κ_{rb}	0	1/3	2/3	1	0
κ_{bs}	0	0	0	0	1

ApplyPheromoneUpdate(cf,bs_update,$\mathcal{T}$,π^{ib},π^{rb},π^{bs}): Our algorithm potentially uses three different solutions for updating the pheromone values: (i) the iteration-best solution π^{ib}, (ii) the restart-best solution π^{rb} and, (iii) the best-so-far solution π^{bs}. Their respective contribution to the update depends on the convergence factor cf, which provides an estimate about the state of convergence of the system. To perform the update, first an update value ξ_{ij} for each pheromone value $\tau_{ij} \in \mathcal{T}$ is computed: $\xi_e := \kappa_{ib} \cdot \delta(\pi^{ib}, i, j) + \kappa_{rb} \cdot \delta(\pi^{rb}, i, j) + \kappa_{bs} \cdot \delta(\pi^{bs}, i, j)$, where κ_{ib} is the weight of π^{ib}, κ_{rb} the weight of π^{rb}, and κ_{bs} the weight of π^{bs} such that $\kappa_{ib} + \kappa_{rb} + \kappa_{bs} = 1.0$. For $i, j = 1, \ldots, n$, the δ-function is the characteristic function, that is, $\delta(\pi, i, j) = 1$ if job j is scheduled directly after job i in permutation π, and $\delta(\pi, i, j) = 0$ otherwise. Similarly, when $i = 0$ and $j = 1, \ldots, n$, $\delta(\pi, 0, j) = 1$ in case j is the first job in permutation π, and $\delta(\pi, 0, j) = 0$ otherwise. Then, the following update rule is applied to all pheromone values:

$$\tau_{ij} = \min \left\{ \max \{ \tau_{\min}, \tau_{ij} + \rho \cdot (\xi_{ij} - \tau_{ij}) \}, \tau_{\max} \right\} \quad ,$$

where $\rho \in (0, 1]$ is the learning rate, set to 0.1. The upper and lower bounds $\tau_{\max} = 0.999$ and $\tau_{\min} = 0.001$ keep the pheromone values always in the range $(\tau_{\min}, \tau_{\max})$, thus preventing the algorithm from converging to a solution. After tuning, the values for κ_{ib}, κ_{rb} and κ_{bs} are chosen as shown in Table 1.

ComputeConvergenceFactor($\mathcal{T}$): This function computes, at each iteration, the convergence factor as

$$cf = 2 \times \left(\left(\frac{\sum_{\tau_{ij} \in \mathcal{T}} \max(\tau_{\max} - \tau_{ij}, \tau_{ij} - \tau_{\min}))}{|\mathcal{T}| \times \tau_{\max} - \tau_{\min}} \right) - 0.5 \right) \tag{2}$$

The convergence factor cf can therefore only assume values between 0 and 1. The closer cf is to 1, the higher is the probability to produce the same solution over and over again.

4   Beam-$\mathcal{MMAS}$

In the following we explain how to obtain a Beam-ACO version from $\mathcal{MMAS}$. The resulting algorithm will henceforth be denoted by Beam-$\mathcal{MMAS}$. In general, Beam-ACO algorithms are hybrids between ACO and beam search [3]. Beam search is an incomplete tree search method [15]. It maintains a search

Algorithm 2. Procedure ProbabilisticBeamSearch(θ, μ)

1: $B_0 = \{\pi = ()\}$
2: $t = 0$
3: **while** $t < n$ **and** $|B_t| > 0$ **do**
4:     $C = \text{ProduceChildren}(B_t)$
5:     **for** $k = 1, \ldots, \min\{\lfloor \mu \cdot \theta \rfloor, |C|\}$ **do**
6:         $\langle \pi, j \rangle = \text{ChooseFrom}(C)$
7:         $C = C \setminus \langle \pi, j \rangle$
8:         $B_{t+1} = B_{t+1} \cup \langle \pi, j \rangle$
9:     **end for**
10:     $B_{t+1} = \text{Reduce}(B_{t+1}, \theta)$
11: **end while**
12: OUTPUT: **if** $t = n$ **then** $\text{argmin}\{f(\pi) \mid \pi \in B_{n-1}\}$ **else** NULL

---

front consisting of a fixed number of θ (partial) candidate solutions. Hereby, θ is called the *beam width*. The search progresses by generating a fixed number of extensions from the current partial solutions and by selecting the θ best extensions with respect to a Greedy function as the new set of open nodes. A cost estimate is used for this purpose, derived from the cost of the partial solution constructed so far and a forward estimate for the remaining components. The quality of this estimate is crucial to the success of beam search.

Beam-ACO algorithms are based on the core idea of using pheromone information for selecting probabilistically the most promising extensions of partial solutions in the search front. This is done in an iterative way. As in ACO, the pheromone values corresponding to good solutions (paths in the search tree) are reinforced at the end of each iteration and guide the search in subsequent iterations. In other words, Beam-ACO algorithms guide basic beam search using a reinforcement learning component in the style of ACO.

Beam-$\mathcal{MMAS}$ is obtained from Alg. 1 by replacing lines 5-10 with $\pi^{ib} = $ ProbabilisticBeamSearch(θ, μ), which calls the probabilistic beam search procedure as shown in Alg. 2. At the start of the procedure the beam only contains an empty permutation, that is $B_0 = \{\pi = ()\}$. At each iteration $t \geq 0$, the algorithm produces the set of all possible children C of the partial permutations that form part of the current beam B_t (see line 4). The extension of a partial permutation $\pi \in B_t$ with job j is henceforth denoted by $\langle \pi, j \rangle$. Note that extending $\pi \in B_t$ with j means placing j at position $t + 1$ of π. At each iteration, at most $\lfloor \mu \cdot \theta \rfloor$ candidate extensions are selected from C by means of the procedure ChooseFrom(C) to form the new beam B_{t+1}. At the end of each step, the new beam B_{t+1} is reduced by means of the procedure Reduce in case it contains more than θ partial solutions. The procedure stops when either B_t is empty, which means that no feasible extensions of the partial permutations in B_{t-1} could be found, or when $t = n - 1$, which means that all permutations in B_{n-1} are complete. In the first case, the algorithm returns NULL, while in the second case the algorithm returns the solution with the best makespan. See Fig. 1(a) for an example of the solution construction.

Procedure ChooseFrom(C) uses the following probabilities for choosing a candidate extension $\langle \pi, j \rangle$ from C:

$$\mathbf{p}(\langle \pi, j \rangle) = \frac{\tau(\langle \pi, j \rangle) \cdot \nu(\langle \pi, j \rangle)^{-1}}{\sum\limits_{\langle \pi', k \rangle \in C} \tau(\langle \pi', k \rangle) \cdot \nu(\langle \pi', k \rangle)^{-1}} \ , \tag{3}$$

where $\tau(\langle \pi, j \rangle)$ corresponds to the pheromone value $\tau_{\pi_t j} \in \mathcal{T}$. The greedy function $\nu(\langle \pi, j \rangle)$ assigns a heuristic value to each candidate extension $\langle \pi, j \rangle$. In principle, we could use the earliest due date heuristic as in Eq. 1 for this purpose. However, when comparing two extensions $\langle \pi, j \rangle \in C$ and $\langle \pi', k \rangle \in C$, their respective earliest due dates might be misleading in case $\pi \neq \pi'$. As in [12], we solved this problem by defining the greedy function $\nu()$ as the sum of the ranks of the earliest due date values that result from the complete construction of the partial solution corresponding to $\langle \pi, j \rangle$. For more information on this procedure see [12].

Finally, the application of procedure Reduce(B_t) removes the worst $\max\{|B_t| - \theta, 0\}$ partial solutions from B_t. For this purpose one usually uses a lower bound for evaluating partial solutions. In fact, this lower bound is generally critical to the performance of the algorithm. However, in the case of the SMJS problem it is surprisingly difficult to find a lower bound that can be efficiently computed. In pilot experiments we discarded several obvious choices, including the minimum setup time estimate and the assignment problem relaxation [7,9]. Due to the lack of an efficient lower bound, we therefore use *stochastic sampling* for evaluating partial solutions. More specifically, for each partial solution a number N^s of complete solutions is sampled. This is done as follows. Starting from the respective partial solution, the solution construction mechanism of $\mathcal{MMAS}$ is used to complete the partial solution in potentially N^s different ways. The only difference to the solution construction mechanism of $\mathcal{MMAS}$ is that when encountering an infeasible partial solution the solution construction is not discarded. Instead, the partial solution is completed, even though it is infeasible. The value of the best one of the N^s samples is used as a measure of the goodness of the respective partial solution. Two measures are considered for comparing different samples: the number of constraint violations, and the makespan. The sample with the lowest number of constraint violations is considered best, and ties are broken with the makespan.

5   Adding CP to $\mathcal{MMAS}$ and Beam-$\mathcal{MMAS}$

Constraint programming (CP) is a paradigm that extends the idea of variables and their domains by attaching restrictions to the domains [13]. A *constraint solver* is used to analyse and to automatically enforce restrictions that the program has requested. The solver provides at least two services to the program: (1) It analyses whether a new restriction is compatible with the already existing ones and signals this as success or failure when a new constraint is added. (2) It

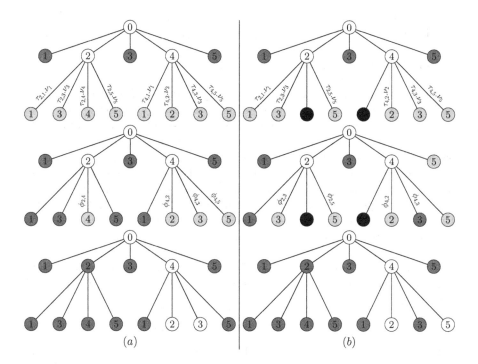

Fig. 1. Partial construction of a solution using Beam-$\mathcal{MMAS}$ and CP-Beam-$\mathcal{MMAS}$. (a) Beam-$\mathcal{MMAS}$: the search proceeds in two stages in order to select subsequent jobs at the next level of the tree. First $\mu \times \theta$ candidates (light gray-filled nodes) are selected using the product of the pheromone information (τ) and rank of a job based on its due time (ν). In this example, $\theta = 2$ and $\mu = 2$. Next, θ solutions are selected using the estimate derived from stochastic sampling ϕ (job 2 and job 3 followed by job 4 in this example, white-filled nodes) and the search proceeds to the next level of the search tree. (b) CP-Beam-$\mathcal{MMAS}$: the search proceeds similar to the Beam-$\mathcal{MMAS}$ method except that infeasible solutions are ruled out by CP at the first stage (black-filled nodes). As this example shows, different solutions may be obtained compared with Beam-$\mathcal{MMAS}$ as a result.

automatically reduces the domains of constraint variables according to the explicit and implicit restrictions. The program can query the solver for the domain of a variable and obtain the set of values that have not been ruled out for this variable. An attempt to assign a concrete value to a variable is called a *labeling* step. Such a tentative assignment may be rejected by the constraint solver (because it is not consistent with the constraints) or it may be accepted and trigger further propagation of constraints. For further details of the CP paradigm the interested reader is referred to [13].

Integrating CP into ACO algorithms can be done by letting the constraint solver keep track of possible extensions to partial solutions. The basic idea is to use constraint variables for the decision variables. We can then use a constraint solver in combination with a constraint model that captures the pertinent relationships between the decision variables to automatically keep track of the feasibility of (partial) solutions. An integration of CP with ant colony system, CP-ACS, has been described in [14] and shown to be effective for the SMJS problem. We use the same CP model in this study.

The CP model. The model maintains two sets of variables. The first set are those variables that were defined in the problem description in Section 1, that is, the positions π_i of permutation π, $\hat{S}$ and $\hat{E}$. A second set of variables from the CP side are introduced which define a stronger model that enhances propagation. Now for the permutation π, auxiliary variables corresponding to the constants are defined: $P = \{p_1, ..., p_n\}$, release times $R = \{r_1, ..., r_n\}$, and due times $D = \{d_1, ..., d_n\}$. Note here that an index i of any of these variables refers to the job at position i in the permutation under construction, i.e., π_i. Additional variables for setup times $ST = \{st_2, ..., st_n\}$ are defined where the index refers to changeover times between consecutive jobs. For example, st_2 refers to the setup time from job π_1 to job π_2. Hence there are only $n-1$ such variables. Variables for the start times ($S = \{s_1...s_n\}$) and end times ($E = \{e_1...e_n\}$) corresponding to $\hat{S}$ and $\hat{E}$ are also defined. Now we can state the CP model. The jobs are first constrained to be unique:

$$\forall i: \quad \pi_i \in \{1, \ldots, n\} \wedge distinct(\pi_i)$$

Strong propagation is then achieved by combining these models by coupling the data and introducing constraints. The data are coupled with the following constraints:[2]

$\forall i: \qquad e_i = s_i + p_i \ \wedge \ s_i \geq r_i \ \wedge \ e_i \leq d_i$

$\forall i > 1: \quad s_i \geq e_{i-1} + st_i \ \wedge \ st_i \in \{\overline{st_{jk}} \mid j, k \in \{1, \ldots, n\}\}$

$\forall i: \qquad d_i \in \{\overline{d_j} \mid j \in 1, \ldots, n\} \wedge r_i \in \{\overline{r_j} \mid j \in 1, \ldots, n\} \wedge p_i \in \{\overline{p_j} \mid j \in 1, \ldots, n\}$

Once the id of the ith job of the permutation is known the following data and variables can be bound:

$\forall i > 1, k, l: \quad \pi_{i-1} = k \ \wedge \ \pi_i = l \Rightarrow st_i = \overline{st_{kl}}$

$\forall i, k: \qquad \pi_i = k \Rightarrow d_i = \overline{d_k} \wedge \ r_i = \overline{r_k} \ \wedge \ p_i = \overline{p_k} \ \wedge \ s_i = \min(\overline{r_k}, e_{i-1} + st_i)$

$\forall i, k: \qquad e_i > d_k \Rightarrow \pi_i \neq k$

High level constraints can be used to achieve efficient propagation and in our case we use *cumulatives(.)*, a scheduling constraint in GECODE [18] which is initialized with all the data and interfaces with the model via $\hat{S}$ and $\hat{E}$. For cross propagation between the two models, the start, end and job id variables are coupled:

$$\forall i, k: \quad \pi_i = k \Rightarrow s_i = \hat{s}_k \ \wedge \ e_i = \hat{e}_k$$

$$\forall i, k: \quad s_k > \hat{s}_i \vee \hat{s}_i < s_k \Rightarrow \pi_i \neq k$$

---

[2] Note that the bars on top labels imply constants and d_i is not the same as $\overline{d_i}$.

Algorithm 3. Construct solutions using CP

1: $i \leftarrow 0$, *feasible* $\leftarrow$ *true*
2: **while** $i \leq n$ & feasible **do**
3:    $i \leftarrow i + 1$
4:    **repeat**
5:      $D = \text{domain}(\pi_i)$
6:      $j = \text{select\_job}(D, \tau)$
7:      **if** $i = 1$ **then**
8:        *feasible* $= \text{post}(\pi_i \leftarrow j \wedge \hat{s}_j \leftarrow 0)$
9:      **else**
10:       *feasible* $= \text{post}(\pi_i \leftarrow j \wedge \hat{s}_j \leftarrow max(\overline{r_j}, \hat{e}_j + \overline{st_{\pi_{i-1},j}}))$
11:      **end if**
12:      **if** not(*feasible*) **then** post($\pi_i \neq j$)
13:    **until** $D \neq \emptyset \vee$ *feasible*
14: **end while**
15: **if** *feasible* **then** return π **else** return NULL

CP can now easily be integrated into the solution construction of $\mathcal{MMAS}$ (see Alg. 3). The algorithm constructs a permutation (π) using the pheromones and CP. This is done in lines 2–14 where the next job is only selected from a feasible domain. D is the candidate list of jobs available for π_i. If it is not possible to schedule the remaining jobs to obtain a feasible solution, the partial candidate solution is discarded without replacement. Effectively, only feasible candidate solutions will be constructed.

The integration of CP into Beam-$\mathcal{MMAS}$ is very similar. In particular, CP is integrated into the phase of Beam-$\mathcal{MMAS}$ which produces possible extensions to the current beam front (line 4 of Alg. 2). The purpose is to restrict the construction of candidate solutions to those that are compatible with the problem constraints. The algorithm implemented here selects all feasible children and ranks them based on their due times as descried earlier.

6   Experiments and Results

We implemented CP-$\mathcal{MMAS}$ and CP-Beam-$\mathcal{MMAS}$ in C++ and used GCC 4.3.2 for compiling the software. The CP code was implemented using the GECODE 2.1.1 solver library.[3] Additionally we re-implemented CP-ACS (originally proposed in [14]), because—for three reasons—it was hard to compare directly to the results obtained by Meyer & Ernst in [14]. First, in [14] labeling steps were used as a measure for the algorithms' termination. Here we instead consider CPU runtime which is a more appropriate measure in practical settings. Secondly, the constraint solver used in the previous study is different to the one used here, and differences between solvers in terms of propagation can be expected. Thirdly, this study also examines the algorithms on data not tested in the previous study.

---

[3] GECODE is available from `http://www.gecode.org/`, accessed on May 20, 2009.

All experiments were run on the Monash Sun Grid and the Enterprisegrid[4] using the parameter sweep application and grid middleware broker Nimrod/G [1]. Each algorithm was applied for 30 times to each considered problem instance, allowing a running time of 10 hours for each application.

6.1   Problem Instances

The problem instances used in this paper include those used in [14] and additional ones from [2]. Individual files are labeled $xn.i$ where $x =$ label, $n =$ instances size and $i =$ index. For example, $W20.1$ is the first file with 20 jobs. The first nine instances are real data taken from wine bottling in the Australian wine industry. There are no release times for this data and the instances are more constrained with decreasing index.

The second set of instances are for the asymmetric traveling salesman problem with time windows and are taken from the literature [2]. The index refers to the number of time windows that have been removed. For example, the RBG problems are identical except that $RBG27.a.15$ has time windows for 15 jobs. Hence, a larger index indicates a more tightly constrained instance.

The third set of instances are selected from the same source as above (see [2]) and aim to demonstrate the algorithms' consistency across a variety of instances. Due to time and resource constraints, instances with a maximum of 152 jobs were considered. Additionally, results for instances with a small number of jobs ($<$ 27) are not reported since the performance of all three algorithms is very similar, nearly always finding an optimal solution.

6.2   Parameter Settings

In the following we give a summary of the parameters settings that were chosen based on past experiments [14,3] and after tuning by hand. In CP-$\mathcal{MMAS}$ and CP-ACS the number of ants was set to 10 per iteration (based on [14]), the learning rate was set to $\rho = 0.01$ (tuned by hand). In $\mathcal{MMAS}$ the upper bound for the pheromone levels was set to $\tau_{max} = 0.999$ and the lower bound to $\tau_{max} = 0.001$.

In CP-Beam-$\mathcal{MMAS}$ we used $\theta = 10$, $\mu = 2.0$ and $N^s = 20$. The choice of $\theta = 10$ is motivated by the fact that CP-$\mathcal{MMAS}$ and CP-ACS use this number of ants at each iteration. The setting of μ was adopted from past experiments with a similar problem [3]. N^s was chosen based on initial experiments.

6.3   Comparison

The results of the three tested algorithms are shown in Table 2, which has the following format. The first column contains the instance identifier. Then, for each

---

[4] The Nimrod project has been funded by the Australian Research Council and a number of Australian Government agencies, and was initially developed by the Distributed Systems Technology CRC.

algorithm we show four different columns. The first one of these four columns presents the best solution found across 30 runs (*best*), the second one shows the mean of the best solutions across all 30 runs (*mean*), the third one contains the associated standard deviations (*sd*), and the fourth one the number of times that the algorithm failed to find a feasible solution (*fail*) out of the 30 runs. In the three columns with heading *best* a value is marked in boldface in case no other algorithm found a better solution and at least one algorithm was not able to match this solution value. Similarly, in the three columns with heading *mean* a value is marked in boldface and italic if no other algorithm has a better mean and at least one algorithm has a worse mean.

First, comparing CP-ACS with the original version presented in [14] shows that our re-implementation performs competitively. Meyer & Ernst use labeling steps as the termination criteria, which results in the fact that the algorithm was run significantly longer than 10 hours for large problems. Despite the restricted time limit our re-implementation is competitive and is significantly worse than the original version only on problem instances W30.1 and RBG027.a.15. However, it is superior to the original version on several other problems (e.g., W30.2). This can be verified by comparing to the results provided in [14].

Concerning the comparison between CP-ACS and CP-$\mathcal{MMAS}$, the former outperforms the latter (sometimes by a large margin) in terms of solution quality when feasible solutions are found. This may be explained by a certain amount of determinism in the solution construction of ACS. However, the failure rate of CP-ACS is very high for instances with 48 or more jobs and this algorithm therefore does not scale well. In contrast, CP-$\mathcal{MMAS}$ does not have such failure issues. In fact, CP-$\mathcal{MMAS}$ is the only algorithm that never fails in finding a feasible solution.

In the following we delve into a comparison of all three algorithms separated for the three instance sets. For the small instances (< 25 jobs) all the algorithms perform equally well except for instance W20.1 where CP-$\mathcal{MMAS}$ performs poorly. In general, for the first set of instances, CP-Beam-$\mathcal{MMAS}$ is the best performing algorithm on the medium-large instances. Even when outperformed, this algorithm always finds solutions close to the best. On the second set of instances the algorithms perform equally well except for RBG27.a.3 where CP-$\mathcal{MMAS}$performs best and RBG27.a.15 where CP-Beam-$\mathcal{MMAS}$ is the best by a large margin. On the third set of instances CP-Beam-$\mathcal{MMAS}$ consistently outperforms the other two algorithms. Again, when the best or best average results are not obtained, CP-Beam-$\mathcal{MMAS}$ always finds solutions whose costs are very close to the best. Furthermore, the algorithm is very robust, finding for many instances always the same best solution value across all 30 runs.

Considering feasibility, as mentioned before, CP-ACS struggles on some of the largest instances, always failing on some (e.g. RBG050b) and nearly always failing on others (all instances of size greater than 67). While CP-$\mathcal{MMAS}$ has no problems in this respect, CP-Beam-$\mathcal{MMAS}$ fails in four out of 90 runs on three instances, demonstrating also a very low failure rate.

Table 2. Results of CP-ACS, CP-$\mathcal{MMAS}$ and CP-Beam-$\mathcal{MMAS}$ for all considered instances

Instance	CP-ACS				CP-$\mathcal{MMAS}$				CP-Beam-$\mathcal{MMAS}$			
	best	mean	sd	fail	best	mean	sd	fail	best	mean	sd	fail
W8.1	8321	8321.0	0.0	0	8321	8321.0	0.0	0	8321	8321.0	0.0	0
W8.2	5818	5818.0	0.0	0	5818	5818.0	0.0	0	5818	5818.0	0.0	0
W8.3	4245	4245.0	0.0	0	4245	4245.0	0.0	0	4245	4245.0	0.0	0
W20.1	**8504**	8542.8	43.8	0	8914	9056.8	86.0	0	**8504**	*8504.0*	0.0	0
W20.2	5062	5076.0	14.0	0	5062	*5062.0*	0.0	0	5062	5068.0	11.3	0
W20.3	4312	4329.7	20.5	0	4312	*4312.0*	0.0	0	4312	4323.7	15.5	0
W30.1	**8127**	8630.0	286.9	0	8887	9068.5	127.7	0	8172	*8334.5*	65.5	0
W30.2	4542	*4641.2*	52.3	0	4682	4839.7	156.8	0	**4527**	4666.3	51.9	0
W30.3	4203	4256.2	49.3	0	4288	4364.2	48.4	0	**4128**	*4217.8*	35.5	0
RBG10.a	3840	3840.0	0.0	0	3840	3840.0	0.0	0	3840	3840.0	0.0	0
RBG16.a	2596	2596.0	0.0	0	2596	2596.0	0.0	0	2596	2596.0	0.0	0
RBG16.b	2094	2094.0	0.0	0	2094	2094.0	0.0	0	2094	2094.0	0.0	0
RBG21.9	4481	4481.0	3.4	0	4481	4481.0	0.0	0	4481	4481.0	0.0	0
RBG27.a.3	**927**	940.7	14.0	0	**927**	*927.1*	0.3	0	940	958.5	13.7	0
RBG27.a.15	1336	1396.6	28.8	0	1500	1543.6	18.5	0	**1068**	*1095.9*	19.3	0
RBG27.a.27	1076	1076.0	0.0	0	1076	1076.0	0.0	0	1076	1076.0	0.0	0
BRI17.a.3	1003	1131.1	203.2	0	1003	*1003.0*	0.0	0	1003	*1003.0*	0.0	0
BRI17.a.10	1031	1130.6	154.6	0	1031	*1031.0*	0.0	0	1031	*1031.0*	0.0	0
BRI17.a.17	1057	1057.0	0.0	0	1057	1057.0	0.0	0	1057	1057.0	0.0	0
RBG027a	**5093**	5135.4	23.0	0	5132	5177.7	18.5	0	**5093**	*5093.0*	0.0	0
RBG031a	3498	3498.0	0.0	0	3498	3498.0	0.0	0	3498	3498.0	0.0	0
RBG033a	3757	3757.0	0.0	0	3757	3757.0	0.0	0	3757	3757.0	0.0	0
RBG034a	3314	3323.9	39.8	0	3314	3362.0	31.1	0	3314	*3314.0*	0.0	0
RBG035a	3388	3413.3	35.1	0	3388	3446.1	44.4	0	3388	*3388.0*	0.0	0
RBG035a.2	3325	3325.0	0.0	0	3325	3325.0	0.0	0	3325	3325.0	0.0	0
RBG038a	5699	5822.0	95.2	0	5699	5914.9	83.5	0	5699	*5699.0*	0.0	0
RBG040a	5679	5695.3	28.5	0	5679	5680.7	5.7	0	5679	*5679.0*	0.0	0
RBG041a	3793	3811.8	35.9	0	3793	3906.3	89.4	0	3793	*3793.0*	0.0	0
RBG042a	**3295**	3399.6	59.7	0	3363	3491.2	71.1	0	3296	*3339.0*	20.6	0
RBG048a	9836	9836.0*	-	29	9856	10019.3	89.3	0	**9799**	9876.6	44.8	1
RBG049a	13257	13296.0	55.2	28	13257	13401.1	72.1	0	13257	*13264.1*	12.9	0
RBG050a	12050	12066.2	37.2	0	12050	12050.9	5.1	0	12050	*12050.0*	0.0	0
RBG050b	-	-	-	30	**12039**	12155.4	69.7	0	12044	*12126.6*	33.6	0
RBG050c	-	-	-	30	11027	11115.1	65.7	0	**10985**	*11015.3*	22.6	1
RBG055a	6929	6990.7	58.6	2	6929	7045.5	64.3	0	6929	*6929.0*	0.0	0
RBG067a	**10331**	10460.1	68.5	21	10368	10485.0	65.5	0	**10331**	*10331.0*	0.0	0
RBG086a	16899	17062.1	94.0	6	16899	17028.8	85.1	0	16899	*16899.0*	0.0	0
RBG092a	12501	12516.3	26.6	27	12501	12530.1	35.8	0	12501	*12501.0*	0.0	0
RBG125a	14296	14399.0	105.1	27	14265	14383.2	98.5	0	**14214**	*14232.2*	25.2	0
RBG132	18524	18592.3	54.9	26	18524	18594.6	74.5	0	18524	*18524.0*	0.0	0
RBG132.2	**18524**	18620.3	110.5	24	18535	18764.5	96.6	0	18524	*18528.2*	14.1	0
RBG152	17455	17455.0	-	29	17455	17455.0	0.0	0	17455	17455.0	0.0	0
RBG152.2	17455	17614.3	156.6	27	17455	17505.8	64.3	0	17455	*17455.0*	0.0	2

* This value is not marked in boldface and italic because it only results from one feasible run.

7 Discussion

The results demonstrate overall that the best algorithm for the SMJS problem is CP-Beam-$\mathcal{MMAS}$. Out of 43 instances examined here Beam-ACO finds the best average solutions for 23 instances and is equal on 15 others. For the remaining five instances CP-ACS obtains the best average twice, however, for RBG048.a only one feasible solution is found in 30 runs. Concerning the three remaining instances (W20.2, W20.3, and RBG27.a.3) CP-$\mathcal{MMAS}$ is the best performing algorithm concerning the average behaviour.

For smaller instances (less than 20 jobs) the algorithms perform equally well as expected. Considering instances of size greater than 20, CP-Beam-$\mathcal{MM}$AS performs better than or equally well to the other algorithms except for two instances. The reason for this improvement can be attributed to the beam search component of CP-Beam-$\mathcal{MM}$AS. First, the time spent in constraint propagation is reduced due to the parallel aspect of the solution construction in beam search. This leads to the fact that CP-Beam-$\mathcal{MM}$AS can spend more time in optimization in contrast to spending time in finding feasibility. Second, this parallel aspect of the solution construction in beam search also allows to discard partial solutions in favor of other partial solutions in different areas of the search tree that appear to be more promising.

In summary, CP-Beam-$\mathcal{MM}$AS is able to make an efficient trade-off between the use of CP for feasibility and the stochastic-heuristic component for finding high quality solutions. Since the CP overhead is large, the CP-ACO variants spend more time propagating constraints compared to CP-Beam-$\mathcal{MM}$AS and are thus less efficient.

8   Conclusion

In this study we have shown that a hybrid algorithm resulting from the combination of Beam-ACO with constraint programming is an effective algorithm for the SMJS problem providing the best known results when optimizing makespan. This method uses CP effectively parallelizing the ACO solution construction and exploiting dependencies between partial solutions to find high quality solutions.

There is certainly room to gain a better understanding of CP-Beam-$\mathcal{MM}$AS's performance. Moreover, more efficient uses of CP within this context may be explored, potentially leading to more time-efficient algorithms. Propagation effectiveness at various stages of variable labeling may be worth examining. Furthermore, components of the algorithm like stochastic sampling can be examined futher. Lopez-Ibanez et al. (see [12]) have analysed stochastic sampling but the success of the method still warrants better understanding. Finally, CP-Beam-$\mathcal{MM}$AS has the potential to be extended to other combinatorial optimization problem types with non-trivial hard constraints where finding feasible solutions is not easy.

Acknowledgements

This work was supported by grant TIN2007-66523 (FORMALISM) of the Spanish government. Moreover, Christian Blum acknowledges support from the *Ramón y Cajal* program of the Spanish Ministry of Science and Innovation.

References

1. Abramson, D., Giddy, J., Kotler, L.: High performance parametric modeling with nimrod/g: Killer application for the global grid? In: International Parallel and Distributed Processing Symposium (IPDPS), pp. 520–528. IEEE Computer Society Press, Los Alamitos (2000)

2. Ascheuer, N., Fischetti, M., Grötschel, M.: Solving the asymmetric travelling salesman problem with time windows by branch-and-cut. Mathematical Programming 90, 475–506 (2001)
3. Blum, C.: Beam-ACO: hybridizing ant colony optimization with beam search: an application to open shop scheduling. Computers and Operations Research 32, 1565–1591 (2005)
4. Blum, C., Blesa, M., Roli, A., Sampels, M. (eds.): Hybrid Metaheuristics: An Emerging Approach to Optimization. Studies in Computational Intelligence, vol. 114. Springer, Heidelberg (2008)
5. Blum, C., Dorigo, M.: The hyper-cube framework for ant colony optimization. IEEE Transactions on Systems, Man, and Cybernetics, Part B: Cybernetics 3, 1161–1172 (2004)
6. Camazine, S., Deneubourg, J.-L., Franks, N.R., Sneyd, J., Theraulaz, G., Bonabeau, E.: Self-Organization in Biological Systems. Princeton University Press, Princeton (2001)
7. Carpaneto, G., Martello, S., Toth, P.: Algorithms and codes for the assignment problem. Annals of Operations Research 13, 193–223 (1988)
8. Coello, C.A.: Theoretical and numerical constraint-handling techniques used with evolutionary algorithms: a survey of the state of the art. Computer Methods in Applied Mechanics and Engineering 191, 1245–1287 (2002)
9. Cormen, T.H., Leiserson, C.E., Rivest, R.L., Stein, C.: Introduction to Algorithms, 2nd edn. MIT Press, Cambridge (2001)
10. Dorigo, M., Stützle, T.: Ant Colony Optimization. MIT Press, Cambridge (2004)
11. Glover, F.W., Kochenberger, G.A. (eds.): *Handbook of Metaheuristics*. International Series in Operations Research & Management Science, vol. 57. Springer, Heidelberg (2003)
12. López-Ibáñez, M., Blum, C., Thiruvady, D., Ernst, A.T., Meyer, B.: Beam-ACO based on stochastic sampling for makespan optimization concerning the TSP with time windows. In: Evolutionary Computation in Combinatorial Optimization. LNCS, vol. 5482, pp. 97–108. Springer, Heidelberg (2009)
13. Marriott, K., Stuckey, P.: Programming With Constraints. MIT Press, Cambridge (1998)
14. Meyer, B., Ernst, A.: Integrating ACO and constraint propagation. In: Dorigo, M., Birattari, M., Blum, C., Gambardella, L.M., Mondada, F., Stützle, T. (eds.) ANTS 2004. LNCS, vol. 3172, pp. 166–177. Springer, Heidelberg (2004)
15. Pinedo, M.L.: Planning and Scheduling in Manufacturing and Services. Springer, New York (2005)
16. Savelsbergh, M.W.P.: Local search in routing problems with time windows. Annals of Operations Research 4, 285–305 (1985)
17. Stützle, T., Hoos, H.H.: Max-min ant system. Future Generation Computer Systems 16, 889–914 (2000)
18. Gecode Team. Gecode: Generic constraint development environment (2008), http://www.gecode.org
19. Valente, J.M.S., Alves, R.A.F.S.: Beam search algorithms for the single machine total weighted tardiness scheduling problem with sequence-dependent setups. Computers and Operations Research 35(7), 2388–2405 (2008)

Multiple Variable Neighborhood Search Enriched with ILP Techniques for the Periodic Vehicle Routing Problem with Time Windows*

Sandro Pirkwieser and Günther R. Raidl

Institute of Computer Graphics and Algorithms
Vienna University of Technology, Vienna, Austria
{pirkwieser,raidl}@ads.tuwien.ac.at

Abstract. In this work we extend a VNS for the periodic vehicle routing problem with time windows (PVRPTW) to a multiple VNS (mVNS) where several VNS instances are applied cooperatively in an intertwined way. The mVNS adaptively allocates VNS instances to promising areas of the search space. Further, an intertwined collaborative cooperation with a generic ILP solver applied on a suitable set covering ILP formulation with this mVNS is proposed, where the mVNS provides the exact method with feasible routes of the actual best solutions, and the ILP solver takes a global view and seeks to determine better feasible route combinations. Experimental results were conducted on newly derived instances and show the advantage of the mVNS as well as of the hybrid approach. The latter yields for almost all instances a statistically significant improvement over solely applying the VNS in a standard way, often requiring less runtime, too.

1 Introduction

The *periodic vehicle routing problem with time windows* (PVRPTW) is a generalized variant of the classical *vehicle routing problem with time windows* (VRPTW) where customers must be served several times in a given planning period instead of only once on a single day. Applications exist in many real-world scenarios as in courier services, grocery distribution, or waste collection.

The PVRPTW is defined on a complete directed graph $G = (V, A)$ with $V = \{0, 1, \ldots, n\}$ being the set of vertices and $A = \{(i, j) \mid i, j \in V, i \neq j\}$ the set of arcs. A planning horizon of t days, referred to by $T = \{1, \ldots, t\}$, is considered. Vertex 0 represents the depot with time window $[e_0, l_0]$ at which are based m vehicles having capacities $Q_1, \ldots, Q_m$ and maximal daily working times $D_1, \ldots, D_m$. Each vertex $i \in V_C$, with $V_C = V \setminus \{0\}$, corresponds to a customer and has associated a demand $q_i \geq 0$, a service duration $d_i \geq 0$, a time window $[e_i, l_i]$, a service frequency f_i, and a set $C_i \subseteq T$ of allowed combinations of visit days. Each arc $(i, j) \in A$ has assigned a travel time (cost) $c_{ij} \geq 0$. The challenge

* This work is supported by the Austrian Science Fund (FWF) under contract number P20342-N13.

M.J. Blesa et al. (Eds.): HM 2009, LNCS 5818, pp. 45–59, 2009.

consists of selecting one visit combination per customer and finding (at most) m vehicle routes on each of the t days on G such that

- each route starts and ends at the depot,
- each customer i belongs to f_i routes over the planning horizon,
- for each vehicle $k = 1, \ldots, m$, the total demand of each route does not exceed capacity limit Q_k, and its daily duration does not exceed the maximal daily working time D_k,
- the service at each customer i begins in the interval $[e_i, l_i]$ and every vehicle leaves the depot and returns to it in the interval $[e_0, l_0]$, and
- the total travel cost of all vehicles is minimized.

Arriving before e_i at a customer i implies a waiting time until this start of the time window (without further cost). Arriving later than l_i is not allowed, i.e. we assume hard time window constraints. In this work, we further assume a homogeneous vehicle fleet with $Q_1, \ldots, Q_m = Q$ and $D_1, \ldots, D_m = D$.

This article introduces a multiple variable neighborhood search (VNS) variant, where several cooperative VNS instances are running in an intertwined way. To obtain even better results, we further consider an *integer linear programming* (ILP) approach based on a set covering formulation and a column generation algorithm and hybridize it in a collaborative way with the multiple VNS.

We refer to related work in Section 2. The variable neighborhood search and its multiple variant are described in Section 3, the ILP formulation in Section 4, and the proposed hybrid method in Section 5. Experimental results are given in Section 6, and Section 7 finishes the work with concluding remarks.

2   Related Work

The PVRPTW was first addressed in [1], where a tabu search algorithm is described for it. In our previous work [2] we suggested a variable neighborhood search (VNS), outperforming the former tabu search. Related VNS metaheuristics exist for the multi-depot VRPTW [3] and the *periodic vehicle routing problem* (PVRP) [4]. Earlier results of our current work, where only a standard VNS (i.e. a single search trajectory) has been combined with a column generation approach and a different cooperation mechanism was used, have been described in [5]. We are not aware of other exact or hybrid methods for the PVRPTW, yet similar PVRPs are dealt with in [6] and [7].

A more general survey of different PVRP variants and solution methods is given in [8]. A similar idea as the one followed in this work was recently applied to a ready-mixed concrete delivery problem [9]. Our work also extends this by highlighting further aspects of this kind of hybridization. In a somewhat related work Danna and Le Pape [10] apply an ILP solver for deriving improved integer solutions during a branch-and-price procedure. For a more general overview on ILP/metaheuristic hybrids we refer to [11].

Although we do not explicitly consider parallelization in this work, our intertwined approach shares features with the replicated parallel VNS variant introduced among other approaches in [12] and also applied in [13].

3   VNS for the PVRPTW

Variable neighborhood search (VNS) [14] is a metaheuristic that applies random steps in neighborhoods with growing size for diversification, referred to as shaking, and uses an embedded local search component for intensification. It has been successfully applied to a wide range of combinatorial optimization problems.In the following we give a rather short overview on our VNS for the PVRPTW as it has been already described in detail in [2].

To smooth the search space, the VNS relaxes the vehicle load, route duration, and time window restrictions and adds penalties corresponding to the excess of these constraints to the cost function. (All three kinds of penalty terms are weighted by a constant factor of 100.) The creation of the initial solution was kept quite simple by selecting a single visit day combination per customer at random and afterwards partitioning the customers at each day into routes. This partitioning is performed by sorting the customers according to the angles they make with the depot—ties are broken using the center of the time windows $(e_i + l_i)/2$—and inserting the customers in this order and a greedy fashion into at most m routes. This insertion is performed in such a way that all but the last routes of each day will comply to the load and duration constraints, while time window constraints might be violated by all routes. The procedure is similar to those introduced in [1].

In the shaking phase we utilize three different neighborhood structures, each with six moves of increasing perturbation size, yielding a total of 18 shaking neighborhoods (i.e. $k_{\max} = 18$) : (i) randomly changing up to six visit combinations with greedy insertion for the new visit days, whereas we also allow reassigning the same visit combination, (ii) moving a random segment of up to six customers of a route to another one on the same day, and (iii) exchanging two random segments of up to six customers between two routes on the same day. In the latter two cases the segments are occasionally reversed. In this work we only consider a fixed shaking neighborhood order.

For intensification we apply the well-known 2-opt intra-route exchange procedure in a best improvement fashion, only considering routes changed during shaking. Additionally each new incumbent solution is subject to a 2-opt* inter-route exchange heuristic [15]. Hereby for each pair of routes of the same day all possible exchanges of the routes' end segments are tried.

To enhance the overall VNS performance not only better solutions are accepted, but sometimes also solutions having a worse objective value. This is done in a systematic way using the Metropolis criterion like in simulated annealing [16]. A linear cooling scheme is used in a way such that the acceptance rate of worse solutions is nearly zero in the last iterations.

3.1   Multiple VNS

We extend the traditional VNS, which only has a single search trajectory, by considering multiple cooperating VNS instances performed in an intertwined way. Thus, our concern here is to investigate the possible benefits of a *sequential cooperative multistart search*. This new VNS variant is denoted as *multiple VNS* (mVNS).

Algorithm 1. Multiple VNS: $\#VNS$ refers to the number of VNS instances, $\#sec$ to the number of sections per VNS instance, and $iter_{max}$ is the total number of allowed iterations.

---

for $i = 1$ **to** $\#VNS$ **do**
    ⌊ initialize VNS[i];
$iter_{sec} \leftarrow \lceil iter_{max}/(\#VNS \cdot \#sec) \rceil$;
for $sec = 1$ **to** $\#sec$ **do**
    for $i = 1$ **to** $\#VNS$ **do**
        ⌊ execute VNS[i] for $iter_{sec}$ iterations;
    $x \leftarrow$ best solution of all VNS instances;
    ⌊ Replace solution of worst VNS instance by x;
Return best solution of all VNS instances;

Although it would be straight-forward to parallelize this approach, parallelization is not the issue we want to focus on here. A somewhat related approach is *replicated parallel VNS* [12,13], in which multiple VNS instances are performed independently in parallel; the overall best solution is finally returned. In this case, the gain in performance is (almost) entirely due to the parallelization. In contrast, we aim at achieving better results within the same total CPU-time as required by a simple VNS run.

The multiple VNS algorithm is shown in Algorithm 1. We initialize each VNS instance independently by performing the method for creating a random solution 100 times and taking the best solution. This way each VNS instance most likely starts with a different initial solution. In the following the VNS instances are executed section-wise by setting an appropriate iteration limit given the total iteration limit and the number of sections. After each block of section-wise executions the actual best solution is determined and replaces the solution of the worst VNS instance. The latter is the cooperative part, where, considered locally, a worse performing VNS is supported by the best one, and seen from a global perspective, the search is intensified in the neighborhood of the so far best solution.

In some sense VNS instances can be said to be adaptively allocated to promising areas of the search space: If a solution is best after one iteration of the outer loop, one additional search trajectory is started from it. If the solution remains the incumbent over further iterations, more VNS instances are restarted from this point and a corresponding stronger intensification takes place. If, however, a new incumbent is found, no further VNS instances will be restarted from the previous one. Of course, in the unlucky event of a very captious local optimum this behavior could lead to a situation where all VNS instances are restarted from the same solution and no further progress is achieved, though this would be no worse than in the single VNS instance case.

In Section 6 we will experimentally compare this multiple VNS to a standard VNS execution and see the advantages w.r.t. final solution quality. The next section introduces a column generation based ILP approach for the PVRPTW problem with which we hybridize the multiple VNS in Section 5 to achieve even better results.

4   Set Covering ILP Model for the PVRPTW

We express the PVRPTW by the following set covering model:

$$\min \sum_{\tau \in T} \sum_{\omega \in \Omega} \gamma_\omega \, \chi_{\omega\tau} \tag{1}$$

$$\text{s.t.} \quad \sum_{r \in C_i} y_{ir} \geq 1 \qquad\qquad \forall i \in V_C \tag{2}$$

$$\sum_{\omega \in \Omega} \chi_{\omega\tau} \leq m \qquad\qquad \forall \tau \in T \tag{3}$$

$$\sum_{\omega \in \Omega} \alpha_{i\omega} \, \chi_{\omega\tau} - \sum_{r \in C_i} \beta_{ir\tau} \, y_{ir} \geq 0 \qquad\qquad \forall i \in V_C; \forall \tau \in T \tag{4}$$

$$y_{ir} \in \{0,1\} \qquad\qquad \forall i \in V_C; \forall r \in C_i \tag{5}$$

$$\chi_{\omega\tau} \in \{0,1\} \qquad\qquad \forall \omega \in \Omega; \forall \tau \in T \tag{6}$$

The set of all feasible routes (satisfying the first, third, and fourth condition from our problem definition in Section 1) visiting a subset of customers is denoted by Ω. Obviously, this set is exponentially large w.r.t. the instance size. For each route $\omega \in \Omega$, let γ_ω be the corresponding costs. We introduce binary variables $\chi_{\omega\tau}$ indicating whether or not route ω is used on day τ, $\forall \omega \in \Omega$, $\tau \in T$. Furthermore, for each customer $i \in V_C$, binary variables y_{ir} indicate whether or not visit combination $r \in C_i$ is chosen. The objective function (1) corresponds to the total costs of all selected routes. Cover constraints (2) guarantee that at least one visit day combination is selected per customer, fleet constraints (3) restrict the number of daily routes to not exceed the available vehicles m, and visit constraints (4) link the routes and the visit combinations, whereas $\alpha_{i\omega}$ and $\beta_{ir\tau}$ are binary constants indicating whether or not route ω visits customer i and if day τ belongs to visit combination $r \in C_i$ of customer i, respectively.

Due to the huge amount of variables it is not possible to directly solve this ILP formulation for instances of practical size. *Column generation*, however, provides a reasonable way to approach such situations [17]: One starts with a small set of initial variables (routes, corresponding to columns in the matrix notation of the ILP) and iteratively extends this set by adding potentially improving variables. In each iteration, the linear programming (LP) relaxation of this reduced problem is (re-)solved in order to finally obtain the solution to the whole LP and thus a lower bound for the original ILP.

This procedure can in principle be combined with branch-and-bound to derive integer solutions, too (and eventually prove their optimality). However, this is another line of research followed by us which turns out to be applicable in a reasonable time to relatively small instances only (roughly about 30 customers

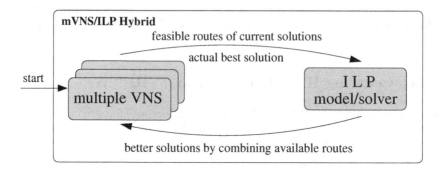

Fig. 1. Information exchange between multiple VNS and the ILP model/solver

at the time of writing). Here, we use the multiple VNS as the sole provider of columns for the set covering model, and restricted ILPs are solved via the general purpose ILP solver CPLEX. This hybrid method is explained in the next section.

5 Hybridizing the Multiple VNS with the ILP Approach

The motivation for the hybrid method is to exploit feasible routes of VNS solutions to derive a new and better solution by applying an ILP solver on the set covering ILP model of Section 4 enriched by these routes. Therefore a pool of solutions' routes is gathered at certain times and provided to the ILP model. This ILP is solved by a branch-and-bound based generic ILP solver, gradually fixing the visit combination (5) and route variables (6). A similar approach was introduced in [9], where the authors highlight the "global view" property of such an exact model. For a set of solutions' routes the ILP solver might be able to derive a more favorable combination and provide a better (less costly) solution in this way. Obviously the potential of the ILP solver depends on the routes contained in the model since it is neither able to alter routes nor to create some on its own. Hence it is crucial to provide (i) a suitable amount of routes, (ii) cost-effective routes, and (iii) diverse enough routes. Adding not enough or only weak routes usually will prevent the ILP solver from finding a better solution at all; on the other hand, a too large set naturally increases the runtime, which might also prevent finding better solutions quickly enough in case a time limit is given. Therefore routes to be added to the model must be carefully selected.

We apply the following hybrid scheme, which can be regarded an intertwined collaborative cooperation [11]; see Figure 1 and Algorithm 2. Concerning the hybridization with the multiple VNS a natural and suitable way is to apply the ILP solver after a block of section-wise executions. This way the number of solutions is given by the number of VNS instances, from which we use the actual best solutions. Due to different search trajectories these solutions' routes are assumed to be diverse enough. Hence, conditions (i)–(iii) are fulfilled. Contrary to [9] we restrict ourselves to the actual best solutions, but we have more VNS instances available. The ILP solver is allotted the same amount of CPU-time than the multiple VNS.

The application of the ILP solver can in some way be regarded as a recombination operator taking into account all available solutions provided by the "population" of the VNS instances. In case a solution is not feasible as a whole, its feasible routes are added anyway, whereas the ILP solver is only applied if at least one feasible solution exists. An additional point of concern is the route injection scheme: It is possible to either add the route for the corresponding day only or for all days. The latter scheme would produce significantly larger ILP models (of factor t) which might yield better solutions at the expense of longer runtimes to solve the ILP model. However, preliminary results as well as results in our previous work [5] suggest to inject routes for the corresponding day only. Further, the ILP solver is always initialized with the current best solution to speed up the process.

If the ILP solver is able to improve on the current best solution this new solution is transferred to the multiple VNS, where as usual the solution of the worst VNS instance is replaced. Before such a transfer eventually over-covered solutions are repaired by choosing exactly one visit combination (the first active) and omitting customers from following routes if they are already covered or do not need to be visited on this day. This over-covering might happen since we use a set covering model. In contrast, a set partitioning model (derived by turning inequalities (2) and (4) into equalities) would yield only feasible solutions but at the same time exclude many potentially improving combinations. Finally, before injecting this solution the previously mentioned 2-opt* improvement procedure is also applied to it. In case routes were altered during this transfer process, corresponding new columns are also added to the ILP model.

There are also two options regarding the lifetime of the routes added to the ILP model: Either we only consider the actual solutions' routes, i.e. they are discarded afterwards, or we keep all inserted routes and the ILP model gradually grows, i.e. realizing a long term memory. However, it is clear that a model of continuously increasing size in general demands more and more computation time to be solved, which could in turn lead to worse solutions when setting a time limit as in our case although the larger search space might also contain better solutions. If routes are kept then the ILP solver is only applied if non-existing routes could be added after a multiple VNS section. Both variants are examined in Section 6.

6   Experimental Results

The algorithms have been implemented in C++, compiled with GCC 4.1 and executed on a 2.83 GHz Intel Core2 Quad Q9550 with 8 GB RAM. We derived new PVRPTW instances from the Solomon VRPTW benchmark instances[1] by evenly assigning the possible visit combinations to the customers at random. We

---

[1] Available at http://web.cba.neu.edu/~msolomon/problems.htm

---

Algorithm 2. Multiple VNS / ILP Hybrid: #*VNS* refers to the number of VNS instances, #*sec* to the number of sections per VNS instance and to the maximal number of ILP solver applications, and $iter_{max}$ is the total number of allowed iterations.

$\Omega' \leftarrow \emptyset$; // start with empty model
for $i = 1$ **to** #*VNS* **do**
  ⌊ initialize VNS[i];
$iter_{sec} \leftarrow \lceil iter_{max}/(\#VNS \cdot \#sec) \rceil$;
for $sec = 1$ **to** #*sec* **do**
  │ $\Omega'_{sec} \leftarrow \emptyset$;
  │ **for** $i = 1$ **to** #*VNS* **do**
  │   │ execute VNS[i] for $iter_{sec}$ iterations;
  │   ⌊ add VNS[i] solutions' routes to Ω'_{sec}; // gather columns
  │ $x^* \leftarrow$ actual best solution;
  │ $\Omega' \leftarrow \Omega' \cup \Omega'_{sec}$; // enrich ILP model
  │ $x \leftarrow$ apply ILP solver on Ω', initialized with x^*;
  ⌊ Replace solution of worst VNS instance by x;
Return best solution of all VNS instances;

---

did so for the first five instances of type random (R), clustered (C), and mixed random and clustered (RC) for a planning horizon of four ($t = 4$) and six days ($t = 6$), denoted by p4 and p6, respectively. For a planning horizon of four days the customers need to be visited either once, two, or four times, for a planning horizon of six days once, two, three, or six times. The number of vehicles m was altered (reduced) in such a way that few or none empty routes occur in feasible solutions, yet it is not too hard to find feasible solutions quite early in the solution process. All instances contain 100 customers and the capacity constraint was left untouched.

For the standard VNS we set an iteration limit of either 10^6 or $2 \cdot 10^6$, an initial temperature of 10 and apply linear cooling every 100 iterations. The multiple VNS as well as the mVNS/ILP hybrid are also allowed 10^6 VNS iterations in total and #*sec* is consistently set to 10 (as determined by preliminary tests); i.e. they apply ten sequences of #*VNS* VNS instances, each one running for $10^6/(\#VNS \cdot 10)$ iterations per section. For solving the ILP model in the mVNS/ILP hybrid we apply the general purpose MIP solver ILOG CPLEX 11.2.

Both mVNS variants were initially run with 5, 10, and 15 VNS instances, these are denoted by mVNS$_{\#VNS,\#sec}$ and mVNS/ILP$_{\#VNS,\#sec}$. Each algorithm setting is run 30 times per instance and we report average results, stating the average travel costs (avg.), corresponding standard deviations (sdv.) and average CPU-times in seconds (t[s]).

The results of the standard VNS are given in Table 1, where we also state the number of vehicles m; this information is omitted in the remaining tables.

As expected, the double amount of iterations also approximately doubles the CPU-time, and for all but instances p4r103 and p6r105 improvements are

Table 1. Results of standard VNS on derived periodic Solomon instances with a planning horizon of four and six days

Instance		VNS (10^6)			VNS ($2 \cdot 10^6$)		
Id	m	avg.	sdv.	t[s]	avg.	sdv.	t[s]
p4r101	14	4141.04	22.52	22.6	4134.47	18.81	44.9
p4r102	13	3759.61	19.43	23.5	3756.77	15.45	46.6
p4r103	10	3191.59	13.56	24.2	3194.75	14.84	47.9
p4r104	7	2613.83	15.76	27.2	2604.74	12.94	54.1
p4r105	11	3697.78	13.84	24.0	3692.96	14.74	47.7
p4c101	10	2910.72	0.53	22.1	2910.53	0.37	44.0
p4c102	8	2963.50	34.01	24.7	2960.70	37.37	48.8
p4c103	7	2806.39	40.13	27.5	2793.80	30.71	54.6
p4c104	7	2481.58	18.14	26.4	2476.27	19.95	52.5
p4c105	8	3025.67	82.95	24.3	2973.57	42.71	48.3
p4rc101	10	4003.77	12.01	25.6	4001.34	14.66	50.7
p4rc102	10	3814.02	19.59	25.3	3798.00	17.47	50.5
p4rc103	8	3500.84	29.40	27.5	3494.06	23.22	55.0
p4rc104	7	3069.41	16.62	27.9	3058.48	18.83	55.4
p4rc105	11	4008.80	25.16	24.5	4001.89	20.06	48.9
p6r101	14	5418.76	10.31	25.9	5417.67	20.27	51.4
p6r102	12	5276.07	23.34	27.0	5261.35	17.65	53.7
p6r103	9	4035.13	28.85	28.8	4014.16	21.37	57.7
p6r104	8	3389.61	16.03	29.5	3380.17	13.35	58.8
p6r105	9	4355.02	27.43	28.3	4355.28	31.80	56.6
p6c101	7	4084.67	36.86	30.1	4076.20	34.75	59.8
p6c102	7	3888.96	20.98	29.7	3876.88	17.27	59.2
p6c103	6	3616.61	44.79	33.8	3583.02	33.04	66.9
p6c104	6	3295.32	18.09	32.8	3291.93	16.76	64.6
p6c105	7	4164.39	64.90	29.9	4139.66	53.95	59.3
p6rc101	10	5846.32	24.69	27.4	5833.84	26.41	54.5
p6rc102	9	5483.15	26.63	28.2	5473.67	24.34	56.1
p6rc103	7	4370.85	25.63	32.2	4360.17	21.22	64.1
p6rc104	7	4147.69	26.00	31.4	4127.46	23.06	63.0
p6rc105	9	5340.30	22.08	28.9	5330.60	19.67	57.7

achieved. In the following the newly introduced algorithm variants will be compared to this "baseline".

Tables 2 and 3 show the results of the mVNS as well as the mVNS/ILP hybrid for the p4 and p6 instances, with the setting of storing all injected routes. At the bottom of each table we additionally state how often the hybrid variant was significantly better or worse than the multiple VNS, as well as how many times both variants were significantly better or worse than the standard VNS variants, whereas we used a Wilcoxon rank sum test with an error level of 5% for testing statistical significance. Best average results are marked bold. Looking

Table 2. Results of mVNS and mVNS/ILP on periodic Solomon instances with a planning horizon of four days

Instance	mVNS$_{5,10}$			mVNS/ILP$_{5,10}$			mVNS$_{10,10}$			mVNS/ILP$_{10,10}$			mVNS$_{15,10}$			mVNS/ILP$_{15,10}$		
	avg.	sdv.	t[s]	avg.	sdv.	t[s]	avg.	sdv.	t[s]	avg.	sdv.	t[s]	avg.	sdv.	t[s]	avg.	sdv.	t[s]
p4r101	4119.88	16.75	22.9	4114.01	13.89	23.3	4121.55	12.42	23.1	4096.09	10.49	26.5	4118.60	12.07	23.2	**4090.09**	5.29	29.0
p4r102	3744.40	6.12	23.8	3744.65	8.15	24.3	3741.79	5.01	24.1	3735.85	4.00	26.6	3742.85	5.03	24.2	**3732.34**	1.94	30.9
p4r103	3186.14	11.19	24.3	3174.98	8.97	25.3	3184.83	9.37	24.8	3171.48	8.44	28.4	3186.70	7.27	24.8	**3165.72**	6.95	33.9
p4r104	2602.15	11.10	27.7	2600.89	10.49	28.9	2602.87	8.63	27.9	**2595.44**	8.20	42.7	2605.36	8.81	28.3	2598.18	10.62	56.0
p4r105	3688.64	12.86	24.1	3681.23	10.05	27.3	3687.94	8.92	24.5	**3679.66**	11.93	48.5	3690.29	8.24	24.6	3686.14	9.16	49.3
p4c101	2910.17	0.26	22.2	2910.24	0.27	22.9	2910.06	0.44	22.6	2909.72	0.67	23.7	2909.91	0.72	22.7	**2909.39**	0.76	24.7
p4c102	2929.76	17.88	24.7	2934.88	23.11	25.2	2930.78	19.63	25.3	2917.46	18.25	28.7	2921.32	20.02	25.4	**2905.16**	14.38	36.4
p4c103	2786.69	31.78	27.8	2774.61	23.01	28.5	2783.91	28.61	28.4	2762.38	18.47	38.6	2777.30	20.18	28.7	**2759.78**	13.82	51.8
p4c104	2465.70	11.65	26.9	2459.19	8.17	27.8	2468.14	10.72	27.0	2459.52	7.60	41.9	2471.83	11.24	27.1	**2454.69**	12.32	49.2
p4c105	2962.17	34.34	24.6	2957.97	36.68	24.8	2952.66	27.78	24.8	2924.55	19.26	31.6	2942.26	29.05	25.0	**2906.69**	15.55	37.6
p4rc101	3988.69	11.61	25.8	3989.75	9.45	26.3	3992.15	10.29	26.1	3978.48	8.95	33.7	3996.60	11.92	26.2	**3974.09**	6.40	46.8
p4rc102	3806.24	17.11	25.6	3796.09	17.08	26.4	3802.65	16.02	26.1	3778.23	14.71	33.9	3801.01	14.74	26.1	**3764.99**	6.76	43.4
p4rc103	3494.62	18.55	27.9	3484.76	18.97	28.5	3500.27	17.08	28.2	3475.95	16.58	31.0	3502.18	23.94	28.3	**3466.99**	12.01	34.7
p4rc104	3042.86	11.52	29.1	3036.39	10.56	30.2	3048.23	11.97	29.2	3036.61	12.43	54.1	3051.19	11.53	29.3	**3031.49**	17.15	57.0
p4rc105	3995.56	13.49	25.1	3991.06	17.18	25.5	3997.55	12.10	25.2	3976.16	8.09	30.1	4000.98	10.76	25.3	**3970.49**	5.67	45.0
significantly better/worse than corresponding mVNS				7×/0×						15×/0×						15×/0×		
significantly better/worse than VNS (10^6)	12×/0×			15×/0×			12×/0×			15×/0×			12×/0×			15×/0×		
significantly better/worse than VNS ($2 \cdot 10^6$)	7×/1×			12×/0×			9×/0×			15×/0×			7×/0×			15×/0×		

Table 3. Results of mVNS and mVNS/ILP on periodic Solomon instances with a planning horizon of six days

Instance	$\text{mVNS}_{5,10}$			$\text{mVNS/ILP}_{5,10}$			$\text{mVNS}_{10,10}$			$\text{mVNS/ILP}_{10,10}$			$\text{mVNS}_{15,10}$			$\text{mVNS/ILP}_{15,10}$		
	avg.	sdv.	t[s]	avg.	sdv.	t[s]	avg.	sdv.	t[s]	avg.	sdv.	t[s]	avg.	sdv.	t[s]	avg.	sdv.	t[s]
p6r101	5402.01	8.78	26.4	5399.76	9.80	27.8	5404.53	7.85	26.6	5390.21	6.10	37.8	5402.38	6.81	26.8	**5385.03**	3.33	49.0
p6r102	5255.09	14.27	27.5	**5244.59**	12.11	32.5	5252.94	11.79	27.9	5249.80	13.81	55.3	5255.77	13.88	28.0	5249.56	14.74	55.6
p6r103	4011.33	20.59	29.4	**3991.46**	12.83	32.5	4006.43	13.54	29.6	3996.78	16.43	58.0	4015.02	14.71	30.0	4008.83	18.72	59.8
p6r104	3376.35	12.41	30.0	3373.36	12.14	34.3	3377.63	10.21	30.4	**3372.81**	9.75	59.1	3382.94	9.40	30.5	3381.98	12.65	60.1
p6r105	4340.76	15.94	28.8	**4337.54**	18.15	31.3	4346.55	15.84	29.2	4337.64	17.22	58.2	4350.98	15.18	29.0	4353.84	18.24	59.4
p6c101	4078.48	27.72	30.6	4060.83	37.72	31.4	**4049.95**	24.47	30.9	4055.21	25.91	59.2	4054.29	20.17	31.1	4050.81	26.17	61.8
p6c102	3867.79	16.38	30.6	3868.96	13.08	32.2	3866.77	13.95	30.6	3861.91	10.82	58.7	**3861.62**	10.81	30.8	3861.86	8.69	62.2
p6c103	3584.92	23.11	34.6	3583.20	29.79	37.5	3589.74	24.60	34.7	**3576.50**	21.80	67.6	3587.16	16.01	35.1	3580.92	18.52	70.3
p6c104	3286.67	15.15	33.5	**3284.07**	15.72	37.5	3287.39	15.14	33.6	3289.73	13.06	65.6	3293.76	17.19	34.1	3291.96	11.87	68.0
p6c105	4124.30	33.03	30.5	4112.89	38.21	31.4	4115.20	27.39	30.8	4119.17	39.47	60.0	4118.48	29.05	30.6	**4104.31**	23.19	61.5
p6rc101	5830.37	13.51	28.0	5831.77	13.67	31.7	5833.97	15.10	28.3	**5821.63**	14.27	56.2	5839.42	14.70	28.5	5833.46	16.74	57.0
p6rc102	5449.68	23.63	29.0	5466.58	25.31	33.0	5468.39	31.07	29.2	**5446.00**	25.84	58.5	5474.09	27.07	29.6	5460.70	30.92	59.0
p6rc103	4366.40	14.85	32.6	**4351.50**	18.34	35.6	4373.90	19.24	33.0	4359.78	18.02	65.1	4373.31	15.87	33.2	4366.85	13.43	66.1
p6rc104	**4130.70**	17.69	31.9	4132.90	21.11	38.3	4146.00	17.74	32.0	4139.72	16.05	63.8	4154.20	18.96	32.3	4149.87	15.64	64.2
p6rc105	5331.66	22.97	29.5	**5321.82**	17.66	32.4	5339.50	14.64	29.5	5329.84	18.28	59.2	5346.52	16.70	30.0	5340.75	15.67	59.7
significantly better/worse than corresponding mVNS				4×/1×						9×/0×						4×/0×		
significantly better/worse than VNS (10^6)	11×/0×			**15×/0×**			10×/0×			13×/0×			7×/0×			10×/0×		
significantly better/worse than VNS ($2 \cdot 10^6$)	5×/0×			8×/0×			4×/3×			**9×/1×**			4×/3×			6×/2×		

Table 4. Results of mVNS and mVNS/ILP on periodic Solomon instances with a planning horizon of six days

Instance	mVNS$_{6,10}$			mVNS/ILP$_{6,10}$			mVNS$_{7,10}$			mVNS/ILP$_{7,10}$			mVNS$_{8,10}$			mVNS/ILP$_{8,10}$			mVNS$_{9,10}$			mVNS/ILP$_{9,10}$		
	avg.	sdv.	t[s]	avg.	sdv.	t[s]	avg.	sdv.	t[s]	avg.	sdv.	t[s]	avg.	sdv.	t[s]	avg.	sdv.	t[s]	avg.	sdv.	t[s]	avg.	sdv.	t[s]
p6r101	5403.42	8.92	26.2	5397.93	7.14	27.9	5402.42	7.19	26.2	5395.88	5.64	28.7	5400.68	7.10	26.3	5392.52	5.13	30.6	5406.09	7.49	26.3	**5391.93**	5.16	33.2
p6r102	5250.62	13.72	27.4	5247.88	14.15	40.2	5251.83	14.73	27.3	5247.96	15.05	50.3	5256.14	14.34	27.4	**5244.00**	13.18	53.7	5251.95	13.12	27.5	5252.02	15.60	54.8
p6r103	4004.61	15.35	29.0	4002.86	16.69	35.3	4009.22	15.93	29.1	3996.57	11.56	39.3	4010.68	10.51	29.1	3998.23	16.13	48.2	4014.87	17.70	29.5	**3996.23**	14.52	54.2
p6r104	3376.94	7.32	29.8	3375.07	10.21	36.4	3381.93	11.76	29.8	3372.69	10.20	44.3	3378.82	10.88	29.8	**3370.82**	11.53	50.0	3383.11	13.68	30.0	3379.34	10.80	56.4
p6r105	4342.18	16.27	28.8	4334.24	15.18	33.5	4344.48	17.94	28.7	**4328.45**	16.34	41.2	4343.22	17.48	28.9	4338.28	14.92	52.3	4344.83	16.69	28.8	4335.15	20.68	55.1
p6c101	4066.66	25.84	30.4	4059.06	24.12	32.6	**4051.95**	31.83	30.4	4056.32	23.01	38.9	4055.49	25.70	30.4	4058.62	33.58	43.4	4059.66	25.49	30.5	4052.32	26.96	51.9
p6c102	3871.16	15.55	30.2	3870.10	14.69	37.5	3868.40	11.06	30.4	**3864.41**	11.47	39.0	3866.25	12.30	30.4	3864.94	12.46	46.4	3868.40	15.05	30.4	3867.13	11.67	54.2
p6c103	3581.14	21.95	34.3	3575.80	24.17	41.0	3593.04	28.02	34.2	3579.95	23.88	48.7	3589.50	22.25	34.5	**3574.98**	24.81	54.6	3579.45	19.34	34.6	3578.30	22.87	61.9
p6c104	3291.55	21.77	33.1	3282.65	16.62	39.6	3287.61	13.54	32.9	3280.17	14.42	46.3	**3278.92**	18.69	33.4	3282.54	15.49	52.8	3288.56	16.01	33.2	3288.17	15.70	60.1
p6c105	4124.76	28.06	30.2	4113.98	27.74	34.8	4110.27	26.88	30.2	4111.19	26.09	40.9	4112.76	25.38	30.2	**4108.25**	27.22	47.8	4111.91	26.67	30.2	4122.30	30.33	57.7
p6rc101	5832.36	14.05	27.7	5823.02	12.67	40.2	5830.79	13.48	28.0	5821.85	13.31	50.0	5834.87	10.53	28.0	**5815.80**	16.78	53.7	5834.19	14.66	28.0	5820.97	17.15	55.4
p6rc102	5464.50	23.69	28.8	5447.06	25.85	37.8	5464.56	25.56	28.8	**5440.55**	27.78	44.1	5461.56	25.45	28.9	5448.11	30.07	53.0	5460.10	22.01	29.0	5442.78	23.82	56.9
p6rc103	4369.68	15.29	32.7	4356.50	15.68	37.5	4365.63	19.79	32.6	**4352.63**	19.56	45.8	4367.29	14.22	32.6	4353.06	19.43	54.2	4368.73	16.70	32.7	4357.15	17.52	60.8
p6rc104	4136.32	17.98	31.6	**4134.89**	20.02	46.7	4146.65	15.45	31.4	4139.70	18.34	55.1	4146.49	14.88	31.6	4136.09	19.17	62.3	4142.93	14.39	31.8	4143.03	17.67	62.9
p6rc105	5326.71	17.63	29.2	5324.04	19.89	37.4	5331.00	20.33	29.4	5323.60	22.41	46.4	5337.35	16.03	29.3	**5318.69**	16.99	54.8	5338.34	18.61	29.5	5325.33	22.58	58.0
significantly better/worse than corresponding mVNS				7×/0×			9×/0×			15×/0×						15×/0×						15×/0×		
significantly better/worse than VNS (10^6)	12×/0×			15×/0×			9×/0×			14×/0×			11×/0×			15×/0×			9×/0×			13×/0×		
significantly better/worse than VNS (2 · 10^6)	2×/2×			10×/1×			5×/2×			11×/1×			6×/1×			12×/1×			6×/1×			8×/1×		

Table 5. Results of mVNS/ILP on periodic Solomon instances with a planning horizon of four and six days when resetting the columns

Instances	mVNS/ILP$_{5,10}$	mVNS/ILP$_{10,10}$	mVNS/ILP$_{15,10}$
significantly better/worse than corresponding mVNS			
p4	$4\times/0\times$	$10\times/0\times$	$14\times/0\times$
p6	$2\times/1\times$	$4\times/0\times$	$9\times/0\times$
significantly better/worse than VNS (10^6)			
p4	$14\times/0\times$	$15\times/0\times$	$14\times/0\times$
p6	$14\times/0\times$	$12\times/0\times$	$11\times/0\times$
significantly better/worse than VNS ($2 \cdot 10^6$)			
p4	$10\times/0\times$	$13\times/0\times$	$13\times/0\times$
p6	$6\times/1\times$	$6\times/1\times$	$8\times/1\times$

at the mVNS it is already quite often better than the standard VNS with 10^6 iterations (up to 80% for p4 instances and 73% for p6 instances) as well as the VNS with $2 \cdot 10^6$ iterations (p4: 60%, p6: 33%), achieving this without additional CPU-time consumption. However, combining the mVNS with ILP techniques in the mVNS/ILP hybrid consistently yields even more satisfying results. In case of mVNS/ILP$_{5,10}$ this is also possible without considerable increase in CPU-time. Although for the variants with #*VNS* set to 10 and 15 the runtime approaches that of the VNS with $2 \cdot 10^6$ iterations (which is per setting the upper limit) for some instances, and thus the comparison to this latter VNS variant is fairer in some sense. For the p4 instances this still results in a 100% success rate, whereas for the p6 instances mVNS/ILP$_{10,10}$ performs best and is 9 times (60%) better and once worse (6.6%) than VNS with $2 \cdot 10^6$ iterations. Looking at the results of the p6 instances, we decided to fine-tune the number of VNS instances, assuming that the performance peak is somewhere between 5 and 10 instances. Therefore we conducted experiments with #*VNS* ranging from 6 to 9; see Table 4. As can be observed the best results are obtained with mVNS/ILP$_{8,10}$, which is, among other settings, always better than the pure mVNS—of course also consuming more CPU-time—and is 12 times (80%) better than the VNS with $2 \cdot 10^6$ iterations, and again only once worse (6.6%). In general, the mVNS/ILP hybrid achieves to a large extent significantly better results than the latter standard VNS yet it still consumes very often less CPU-time.

So far we only considered the strategy to keep all routes in the model once they were added. The results of the hybrid algorithm when resetting the columns after each application of the ILP solver are given in Table 5. Due to space limitations we only state the results of the statistical significance tests. For the p4 instances there is clearly no gain, whereas not storing the routes has the greatest impact when using 15 VNS instances, i.e. when most columns are added. Here, the reduced size of the model leads to more improvements in the limited time. Nevertheless, apart from less runtime in total for obvious reasons, it seems generally better to work with an ILP model of increasing size and hence exploit information from the search trajectory.

7   Conclusions

We extended our previously introduced (standard) VNS for the periodic vehicle routing problem with time windows (PVRPTW) to a multiple VNS (mVNS) where several VNS instances are applied cooperatively in an intertwined way. This mVNS puts emphasis on the so far best solution found within a major iteration by restarting the worst performing VNS instances with it. In this way, mVNS investigates multiple search trajectories from incumbent solutions, and from a global perspective it can be seen to adaptively allocate VNS instances to promising areas of the search space. Further an intertwined cooperative combination of this mVNS and a generic ILP solver applied to a suitable set covering ILP formulation was proposed. The mVNS provides the exact method with feasible routes of the actual best solutions, and the ILP solver takes a global view and seeks to determine better feasible route combinations. For testing we derived new PVRPTW instances with a planning horizon of four and six days from the 100 customer Solomon VRPTW benchmark instances. Experimental results showed the advantages of the mVNS as well as of the hybrid approach, the latter yielding for 80%–100% of all conducted tests a statistically significant improvement over solely applying the VNS in a standard way. It has become evident that keeping the routes (columns) in the model once they were added is beneficial, though one has to keep in mind the longer runtimes of this setting than when considering the actual best solutions' routes only. Nevertheless, even the mVNS/ILP hybrid with this continuously increasing ILP model—clearly performing best of all variants—requires for most of the instances less CPU-time than the standard VNS with more iterations.

As future work we might consider some sort of column management for the ILP to have an additional option in-between resetting the columns or persistent storage. Also of interest for the mVNS would be a special perturbation operator (probably a more destructive shaking) in case the same local optimal solution is injected more than once in a VNS instance. From a practical perspective, dealing with customer demands depending on the visit day would be interesting, too. Last but not least, we want to remark that the general approach described here also might be promising for many other combinatorial optimization problems.

References

1. Cordeau, J.F., Laporte, G., Mercier, A.: A unified tabu search heuristic for vehicle routing problems with time windows. Journal of the Operational Research Society 52, 928–936 (2001)
2. Pirkwieser, S., Raidl, G.R.: A variable neighborhood search for the periodic vehicle routing problem with time windows. In: Prodhon, C., et al. (eds.) Proceedings of the 9th EU/MEeting on Metaheuristics for Logistics and Vehicle Routing, Troyes, France (2008)

3. Polacek, M., Hartl, R.F., Doerner, K., Reimann, M.: A variable neighborhood search for the multi depot vehicle routing problem with time windows. Journal of Heuristics 10, 613–627 (2004)
4. Hemmelmayr, V.C., Doerner, K.F., Hartl, R.F.: A variable neighborhood search heuristic for periodic routing problems. European Journal of Operational Research 195(3), 791–802 (2009)
5. Pirkwieser, S., Raidl, G.R.: Boosting a variable neighborhood search for the periodic vehicle routing problem with time windows by ILP techniques. In: Caserta, M., Voß, S. (eds.) Proceedings of the 8th Metaheuristic International Conference (MIC 2009), Hamburg, Germany (2009)
6. Francis, P., Smilowitz, K., Tzur, M.: The period vehicle routing problem with service choice. Transportation Science 40(4), 439–454 (2006)
7. Mourgaya, M., Vanderbeck, F.: Column generation based heuristic for tactical planning in multi-period vehicle routing. European Journal of Operational Research 183(3), 1028–1041 (2007)
8. Francis, P.M., Smilowitz, K.R., Tzur, M.: The period vehicle routing problem and its extensions. In: Golden, B., et al. (eds.) The Vehicle Routing Problem: Latest Advances and New Challenges, pp. 73–102. Springer, Heidelberg (2008)
9. Schmid, V., Doerner, K.F., Hartl, R.F., Savelsbergh, M.W.P., Stoecher, W.: A hybrid solution approach for ready-mixed concrete delivery. Transportation Science 43(1), 70–85 (2009)
10. Danna, E., Le Pape, C.: Branch-and-price heuristics: A case study on the vehicle routing problem with time windows. In: Desaulniers, G., et al. (eds.) Column Generation, pp. 99–129. Springer, Heidelberg (2005)
11. Raidl, G.R., Puchinger, J.: Combining (integer) linear programming techniques and metaheuristics for combinatorial optimization. In: Blum, C., et al. (eds.) Hybrid Metaheuristics: An Emerging Approach to Optimization. Studies in Computational Intelligence, vol. 114, pp. 31–62. Springer, Heidelberg (2008)
12. García-López, F., Melián-Batista, B., Moreno-Pérez, J.A., Moreno-Vega, J.M.: The parallel variable neighborhood search for the p-median problem. Journal of Heuristics 8(3), 375–388 (2002)
13. Moreno-Pérez, J.A., Hansen, P., Mladenović, N.: Parallel variable neighborhood search. In: Alba, E. (ed.) Parallel Metaheuristics: A New Class of Algorithms, pp. 247–266. John Wiley & Sons, NJ (2005)
14. Hansen, P., Mladenović, N.: Variable neighborhood search. In: Glover, F., Kochenberger, G. (eds.) Handbook of Metaheuristics, pp. 145–184. Kluwer Academic Publishers, Boston (2003)
15. Potvin, J.Y., Rousseau, J.M.: An exchange heuristic for routeing problems with time windows. Journal of the Operational Research Society 46, 1433–1446 (1995)
16. Kirkpatrick, S., Gelatt Jr., C.D., Vecchi, M.P.: Optimization by simulated annealing. Science 220(4598), 671–680 (1983)
17. Desrosiers, J., Lübbecke, M.E.: A primer in column generation. In: Desaulniers, G., et al. (eds.) Column Generation, pp. 1–32. Springer, Heidelberg (2005)

A Hybridization of Electromagnetic-Like Mechanism and Great Deluge for Examination Timetabling Problems

Salwani Abdullah[1], Hamza Turabieh[1], and Barry McCollum[2]

[1] Center for Artificial Intelligence Technology,
Universiti Kebangsaan Malaysia, 43600 Bangi, Selangor, Malaysia
{salwani,hamza}@ftsm.ukm.my
[2] Department of Computer Science, Queen's University Belfast,
Belfast BT7 1NN United Kingdom
b.mccollum@qub.ac.uk

Abstract. In this paper, we present a hybridization of an electromagnetic-like mechanism (EM) and the great deluge (GD) algorithm. This technique can be seen as a dynamic approach as an estimated quality of a new solution and a decay rate are calculated each iteration during the search process. These values are depending on a force value calculated using the EM approach. It is observed that applying these dynamic values help generate high quality solutions. Experimental results on benchmark examination timetabling problems demonstrate the effectiveness of this hybrid EM-GD approach compared with previous available methods. Possible extensions upon this simple approach are also discussed.

Keywords: Hybrid metaheuristic, Electromagnetism-like mechanism, Great deluge, Decay rate, Exam timetabling.

1 Introduction

Examination timetabling problems are very common in schools and universities. Solutions to the problem concerns the allocating a set of exams, into a limited number of timeslots (periods), subjects to a set of constraints. Carter et al. [10] stated that the basic challenge of the examination timetabling problem is to schedule examinations over a limited set of timeslots so as to avoid conflicts and to satisfy a number of side-constraints. In this case, the conflict is referred as hard constraints and side-constraints are referred as soft constraints. Generally accepted hard constraints are (i) there must be enough seating capacities and (ii) no student should be required to sit two examinations at the same time. Solutions that satisfy all hard constraints are often called feasible solution. The most common set of soft constraints as reported in Burke et al. [5] are: students should not be scheduled to sit more than one examination in a day; students should not be scheduled to sit examinations in two consecutive timeslots; each student's examinations should be spread as evenly as possible over the schedule

M.J. Blesa et al. (Eds.): HM 2009, LNCS 5818, pp. 60–72, 2009.
© Springer-Verlag Berlin Heidelberg 2009

and examinations of the same length should be scheduled in the same room. In real world situation, it is, of course, usually impossible to satisfy completely all soft constraints. Therefore efforts are made in minimizing these violations to increase the quality of the solution by calculating the extent to which a solution has violated the set of soft constraints. McCollum et al. [18,17] introduced a new formulation of the problem as part of ITC2007. This more comprehensive description of the problem describes a range of hard and soft constraints found in recent practical problems from within the literature and practical experience. The intention here is to show the effectiveness of our technique on the original Carter datasets (see Carter et al. [12], Burke et al. [6]) before continuing the applications to this recently introduced formulation.

In the past, a wide variety of approaches for solving the examination timetable problem have been described and discussed in the literature. These can be categorized into: sequential methods, cluster methods, constraint-based methods, generalised search (meta-heuristics), multi-criteria approaches, case based reasoning techniques, and hyper-heuristics/self adaptive approaches (Carter and Laporte [11], Petrovic and Burke [20]). These approaches are tested on various examination timetabling datasets that can be downloaded from `http://www.asap.cs.nott.ac.uk/resource/data`. Interested readers can find more details about examination timetabling research in the comprehensive survey paper by Qu et al. [22] and Lewis [16].

The remainder of this paper is as follow; Section 2 provides a review on available hybrid approaches applied to examination timetabling problems; Section 3 provides the necessary information on the formulation of the examination timetabling problem; Section 4 describes the detailed implementation of the electromagnetic metaheuristic along with the neighbourhood structures used. The simulation results are presented in Section 5. Finally, the paper is concluded making comment on the effectiveness of the technique studied and potential future research approaches.

2   Hybrid Approach: Research and Developments

A hybrid approach is one which subsumes two or more methods. The advantage of combining several methods is that potentially the process helps to compensate for the insufficiency of using each type of method in isolation.

Burke et al. [5] developed a memetic algorithm where light and heavy mutations were employed. Hill climbing was used to improve the quality of timetables. The approach was tested on Nottingham dataset (see Qu et al. [22]). Merlot et al.[19] employed constraint programming to produce initial solutions. A simulated annealing approach was used to improve the solution. Subsequently a hill climbing method is employed to further improve the quality of the solutions. The overall hybrid approach was tested on the Carter, University of Melbourne and Nottingham datasets. For details on examination timetabling benchmark data, please refer to Qu et al. [22]. This approach obtained some of the best results reported in the literature. Casey and Thompson [13] investigated a greedy randomized adaptive search procedure (GRASP) approach where the initial solution was generated by a modified saturation degree heuristic. Backtracking was employed in conjunction with a tabu list. A simulated annealing approach (with kempe chain moves) with a high starting temperature and fast cooling scheme was used in the improvement phase. The approach was applied on Carter's dataset and able to obtain competitive results at the time. Burke et al. [7] investigated a knowledge based technique i.e. case-based reasoning as a heuristic selector and tested this on four examination datasets. Different ways of hybridizing the low level graph heuristics were compared for solving Carter's datasets and was shown able to produce good results. Yang and Petrovic [23] employed case-based reasoning to choose graph heuristics to construct initial solutions for subsequent use of the great deluge algorithm. The approach obtained the best results reported in the literature for several Carter's datasets. Abdullah and Burke [1] investigated a hybridization of the very large neighbourhood search approach with local search methods. The approach initially investigated where a very large neighbourhood of solutions using graph based algorithms. The second phase made further improvement by utilizing local search methods (simulated annealing and great deluge individually). Other related works on hybrid approaches applied on university timetabling problems can be found in Côtè et al. [14] which utilised a hybrid multi-objective evolutionary. Qu and Burke [21] investigated hybridization within a graph based hyper-heuristic for university timetabling problems. Qu et al. [22] employed a dynamic hybridisation of different graph colouring heuristics, and Burke et al. [8] employed hybrid variable neighbourhood search for the same instances.

3   Problem Description

The problem description employed as part of this work is adapted from the description presented in Burke et al. [6]. The problem can be stated as follows:

- E_i is a collection of N examinations $(i = 1, \ldots, N)$.
- T is the number of timeslots.
- $C = (c_{ij})_{N \times N}$ is the conflict matrix where each record, denoted by $c_{ij}(i, j \in \{1, \ldots, N\})$, represents the number of students taking exams i and j.
- M is the number of students.
- $t_k(1 \le t_k \le T)$ specifies the assigned timeslots for exam $k(k \in \{1, \ldots, N\})$.

The following hard constraint is considered based on Carter et al. [12]:
no students should be required to sit two examinations simultaneously.
In this problem, we formulate an objective function which tries to spread out exams throughout the exam period (Expression (1)).

$$Min \quad \frac{\sum_{i=1}^{N-1} F(i)}{M} \qquad (1)$$

where:

$$F(i) = \sum_{j=i+1}^{N} c_{ij} \cdot proximity(t_i, t_j) \tag{2}$$

$$proximity(t_i, t_j) = \begin{cases} 2^5/2^{|t_i - t_j|} & \text{if } 1 \le |t_i - t_j| \le 5 \\ 0 & \text{otherwise} \end{cases} \tag{3}$$

subject to:

$$\sum_{i=1}^{N-1} \sum_{j=i+1}^{N} c_{ij} \cdot \lambda(t_i, t_j) = 0 \quad where: \quad \lambda(t_i, t_j) = \begin{cases} 1 & \text{if } t_i = t_j \\ 0 & \text{otherwise} \end{cases} \tag{4}$$

Equation (2) presents the cost for an exam i which is given by the proximity value multiplied by the number of students in conflict. Equation (3) represents a proximity value between two exams (Carter et al. [12]). Equation (4) represents a clash-free requirement so that no student is required to sit two exams at the same time.

4   A Hybridisation of Electromagnetism-Like Mechanism and Great Deluge Algorithm

4.1   Solution Representation

As part of this work, a direct representation is employed. Each population member is represented as a number of genes that contain information about the timeslot and associated exams. Fig.1 shows examples of the genes where t_j is a timeslot ($j \in \{1, \dots, T\}$), e_i is an exam ($i \in \{1, \dots, N\}$). For example; $e_2, e_{11}, e_8, e_7, e_{14}$ are allocated in timeslot t_1, while e_{21}, and e_{19} allocated in t_4.

4.2   Electromagnetic-like Mechanism

Birbil and Fang[3] constructed the electromagnetic-like mechanism drawing upon the attraction-repulsion mechanism of the theory of electromagnetism. Each

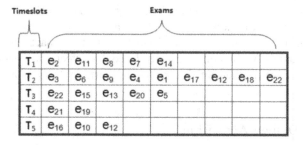

Fig. 1. Solution representation

sample point (timetable in this case) is released to a space as a charged particle whose charge relates to the objective function value. The charge determines the magnitude of attraction or repulsion of the point over the sample population. The better the objective function value, the higher the magnitude of attraction. The attraction directs the points toward better regions, whereas repulsion allows particles to exploit the unvisited regions. In this paper, the electromagnetism-like mechanism starts with a population of randomly generated timetables. The static force between two point charges is directly proportional to the magnitudes of each charge and inversely proportional to the square of the distance between the charges (see Birbil and Fang [3]). In this work, the fixed charge of timetable i is calculated as follows:

$$q^i = exp\left(-T\frac{f(x^i) - f(x^b)}{\sum_{k=1}^{m}(f(x^k) - f(x^b))}\right) \tag{5}$$

where:
q^i: the charge for timetable i.
$f(x^i)$: penalty of timetable i.
$f(x^k)$: penalty of timetable k.
$f(x^b)$: penalty of timetable b (b=best timetable through population).
m: population size.
T: number of timeslots.
The solution quality or charge of each timetable determines the magnitude of an attraction and repulsion effect in the population. A better solution encourages other particles to converge to attractive valleys while a bad solution discourages particles to move toward this region. These particles move along with the total force and so, diversified solutions are generated. The following formulation is the total force of particle i.

$$F_{ij} = \sum_{j\neq i}^{m}\begin{cases}(f(x^j) - f(x^i))\frac{q^i q^j}{\left\|f(x^j)-f(x^i)\right\|^2} & \text{if } f(x^j) < f(x^i) \\ (f(x^i) - f(x^j))\frac{q^i q^j}{\left\|f(x^j)-f(x^i)\right\|^2} & \text{if } f(x^j) \geq f(x^i)\end{cases}, \forall i \tag{6}$$

In general, the process of evaluating the total force is illustrated in Fig.2. Three particles (labeled as 1, 2 and 3) which represent feasible solutions along with their associated objective function values i.e. 30, 8 and 19, respectively. Since particle 1 is worst than particle 3, a repulsive force F_{13} effects on particle 3. Particle 2 is better than particle 3, thus an attractive force F_{23} effects on particle 3 to force attraction-repulsive in different directions, i.e. particle 3 moves along with total force F.

4.3   Great Deluge Algorithm

The great deluge algorithm was introduced by Dueck [15]. It is a local search procedure which has certain similarities with simulated annealing. This approach

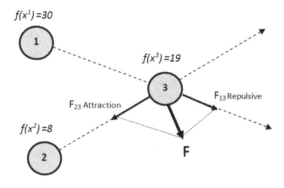

Fig. 2. Attraction- repulsive effect on particle 3

is far less dependent upon parameters than simulated annealing. It needs just two parameters: the amount of computational time that the user wishes to "spend" and an estimate of the quality of solution that a user requires. Apart from accepting a move that improves the solution quality, the great deluge algorithm also accepts a worse solution if the quality of the solution is less (for the case or minimisation) than or equal to some given upper boundary value β (in the paper by Dueck it was called a "level"). In this work, the "level" is initially set to be the objective function value of the initial solution. During its run, the "level" is iteratively lowered by a constant where is a force decay rate (see Fig.3). The great deluge algorithm will be applied on each timetable to reduce the total penalty cost based on the calculated force value.

The pseudo code for the great deluge is presented in Fig.3. In this work, two types of neighbourhood structures have been applied i.e.:   '
N1: Select two exams at random and swap timeslots.
N2: Choose a single exam at random and move to a new random feasible timeslots.

4.4   A Hybrid Approach

The discussion on the hybrid approach is divided into two parts i.e. (1) Initialisation and (2) electromagnetic-like mechanism and great deluge algorithm as shown in Fig.4. In Step 1, the quality of the initial and best solutions are calculated and set together with the number of iteration and level. In Step 2, electromagnetic-like mechanism is implemented to calculate the force for each solution. The force value later will be used in the great deluge algorithm to calculate the decreasing rate (in this paper, we referred as a force decay rate) as shown in Fig.4.

Fig.5 illustrates an example of seven solutions with the current objective function values (penalty). For each solution, firstly, the charge and force have to be calculated and evaluated using equations (5) and (6) respectively. For example,

Calculate estimated quality of every solution, EstimatedQuality = f(Sol$_i$)- F$_i$, where
i = 1 to population size;

Calculate force decay rate, β = *EstimatedQuality/NumOfIte;*

Define neighbourhood (N$_1$ and N$_2$) of Sol$_i$ by randomly assigning exam to a valid
timeslot to generate a new solution called Sol$_i$;*

Calculate f(Sol$_i$);*

if (f(Sol$_i$) < f(Sol$_{best}$)) where Sol$_{best}$ represent the best solution found so far*
     Sol$_i$ ← Sol$_i$;*
     Sol$_{best}$ ← Sol$_i$;*
else
        if (f(Sol$_i$)≤ level)*
        Sol$_i$← Sol$_i$;*
     level = level - β ;

Increase iteration by 1;

Fig. 3. Pseudo code for the great deluge algorithm

Step 1: Initialization:
        Set initial solution as Sol;
        Calculate the initial penalty cost, f(Sol);
        Set best solution, Sol$_{best}$ ← Sol;
        Set total number of iterations, NumOfIte;
        Set initial level: level ← f(Sol);
        Set iteration ← 0;

Step 2: Evaluation:
     ***do while** (iteration < NumOfIte)*
        Calculate total force, F, for each timetable based on
        electromagnetic-like mechanism;
        Apply great deluge algorithm (see Figure 3)
     end do

Fig. 4. Pseudo-code for the hybrid approach

the penalty, charge and force for Sol_1 are 175.2292, 0.1161401 and 0.0769340, respectively. Sol_6 has a charge value = 1 and force = 0. This means that the quality of this timetable (*particle*) is not needed to be reduced in the next iteration. However, for the rest of the solutions, the penalties are attempted to be lowered at least at (penalty - force). Taking Sol_1 as an example, the force (F) is 0.0769340, thus the estimated quality of the solution,*Estimated Quality*, (see Fig.3) will be 175.1522 (i.e. 175.2292 -0.0769340). However, the great deluge algorithm is able to reduce the penalty for each timetable less than the estimated quality. For example, after applying the great deluge, the quality of Sol_1 is 174.4521 (which is less than 175.1522). Note that, this example is taken from our experiment on sta-f-83 dataset.

	Sol $_1$	Sol $_2$	Sol $_3$	Sol $_4$	Sol $_5$	Sol $_6$	Sol $_7$
Penalty	175.2291	175.4206	175.2569	175.2422	175.2094	175.4206	175.2569
Charge	0.1161401	0.10887592	0.11505530	0.11562836	0.1169120	1	0.114928357
Force	0.0769340	0.171878	0.12942298	0.10339403	0.0501987	0	0.1557118

$$\Downarrow$$

Great Deluge

$$\Downarrow$$

	Sol $_1$	Sol $_2$	Sol $_3$	Sol $_4$	Sol $_5$	Sol $_6$	Sol $_7$
Penalty	174.4521	173.6521	173.666	175.000	170.3220	175.4206	169.3254
Charge	0.0814217	0.1164618	0.1157398	0.06372075	0.51669789	0.0775116	0.80701278
Force	0.1151856	0.1147518	0.1409154	0.10927126	0.181958767	0.1219733	0.7840562

Fig. 5. Illustrative example for sta-f-83 with seven solutions

5   Simulation Results

The proposed algorithm was programmed using Matlab and simulations were performed on the Intel Pentium 4 2.33 GHz. In this paper, we considered a standard benchmark examination timetabling problem from Carter et al. [12]. Table 1 shows the parameter for the hybrid algorithm after some preliminary experiments.

Table 2 provides the comparison of our results with the best known results for these benchmark datasets (taken from Qu et al. [22]). The best results out of 5 runs are shown in bold.

Our algorithm produces better results on seven out of eleven datasets. We are particularly interested to compare our results with the other results in the literature that employed a hybrid approach i.e.: Merlot et al. [19] employed constraint programming as initialization for simulated annealing and hill climbing; Casey and Thompson [13] applied GRASP with modified saturation degree

Table 1. Parameter Setting

Parameter	Value
Generation number	10000
Population size	50

Table 2. Comparison Results

Instance	Our approach	Best Known	Authors for best known
car-s-91	**4.46**	4.50	Yang and Petrovic [23]
car-f-92	**3.76**	3.98	Yang and Petrovic [23]
ear-f-83	32.12	**29.3**	Caramia et al. [9]
hec-s-92	9.72	**9.2**	Caramia et al. [9]
kfu-s-93	**12.62**	13.0	Burke et al. [4]
lse-f-91	10.03	**9.6**	Caramia et al. [9]
sta-f-83	**156.94**	157.2	Côtè et al. [14]
tre-s-92	**7.86**	7.9	Burke et al.[5]
uta-s-92	**2.99**	3.14	Yang and Petrovic [23]
ute-s-92	24.9	**24.4**	Caramia et al. [9]
yor-f-83	**34.95**	36.2	Caramia et al. [9], Abdullah et al.[2]

initialization and simulated annealing; Yang and Petrovic [23] employed fuzzy set on selecting hybridizations of great deluge and graph heuristics; Abdullah and Burke [1] employed large neighbourhood search approach with local search methods, Qu and Burke [21] that investigated the hybridization within a graph based hyper-heuristic; Qu et al.[22] employed a dynamic hybridisation of different graph colouring heuristics; and Burke et al.[8] investigated a hybrid variable neighbourhood search.

Table 3 shows the comparison results on the hybrid algorithms as mentioned earlier. Again, the best results out of 5 runs are shown in bold. Note that the value marked "-" indicates that the corresponding problem is not tested.

From Table 3, we can see that our algorithm produces better results on almost all datasets (accept on sta-f-83 and yor-f-83 datasets) when compared against other hybridization methods. Note that Casey and Thompson[13] used different version of datasets (denoted in italic). It is clearly shown that our hybrid approach out performs other hybrid approaches on all of the instances. We believe that with the help of a force decay rate generated from electromagnetism-like mechanism to determine the estimated quality and its subsequent use as a level in the great deluge algorithm manages to reduce the penalty cost of the solution. Furthermore, in our case we applied this improvement hybrid algorithm to all the population rather than a single population. This implementation helps the whole population to converge together because only solutions that are better than the best solution so far or better than the level will be added into the population pool to be used in the next iteration.

Fig.6 shows the convergence of the hybrid approach when applied on sta-f-83 dataset. The x dimension represents the number of iteration while the y dimension represents the penalty cost. Each point shown the graph represents the solution found in each iteration. The curve up and down from the beginning until it becomes stagnant to the end of the search. This algorithm will always accepts best solutions and worse solutions will be accepted as long their quality is better than the level. This is probably because there are more rooms for

Table 3. Comparison Results On Hybrid Approaches

Instance	Our Approach		Merlot et al. (2003)	Casey and Thompson (2003)	Côtè et al. (2005)
	Best	Average			
car-s-91	**4.42**	4.81	5.1	5.4	5.2
car-f-92	**3.76**	3.95	4.3	4.2	4.2
ear-f-83	**32.12**	33.69	35.1	34.2	34.2
hec-s-92	**9.73**	10.10	10.6	10.2	10.2
kfu-s-93	**12.62**	12.97	13.5	14.2	14.2
lse-f-91	**10.03**	10.34	10.5	14.2	11.2
sta-f-83	156.94	157.30	157.3	*134.9*	157.2
tre-s-92	**7.86**	8.2	8.4	8.2	8.2
uta-s-92	**2.99**	3.32	3.5	-	3.2
ute-s-92	**24.9**	25.41	25.1	25.2	25.2
yor-f-83	34.95	36.27	37.4	37.2	36.2

Instance	Abdullah and Burke (2006)	Yang and Petrovic (2005)	Qu and Burke (2008)	Qu et al. (2009)	Burke et al. (2006)
car-s-91	4.1	4.50	5.16	5.11	4.6
car-f-92	4.8	3.93	4.16	4.32	3.9
ear-f-83	36.0	33.70	35.86	35.56	32.8
hec-s-92	10.8	10.83	11.94	11.62	10.0
kfu-s-93	15.2	13.82	14.79	15.18	13.0
lse-f-91	11.9	10.35	11.15	11.32	10.0
sta-f-83	159.0	158.35	159	158.88	156.9
tre-s-92	8.5	7.92	8.6	8.52	7.9
uta-s-92	3.6	3.14	3.59	3.21	3.2
ute-s-92	26.0	25.39	28.3	28	24.8
yor-f-83	36.2	36.35	41.81	40.71	34.9

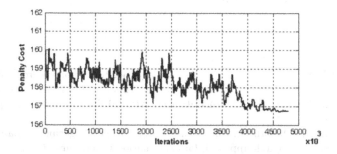

Fig. 6. Sta-f-83 convergence using hybrid approach

improvement at the beginning of the search and towards the end of the search process less improvement is achieved. We believe that this algorithm is able to produce some of best known solutions because the amount of the decreasing rate (based on the force value) that is calculated at every iteration is very small (see Fig.6), thus the level to be decreased is also very small. This helps the algorithm to easily accepts the solutions with small improvement with respect to the objective function. It is also believed that better solutions can be obtained in these experiments with the help of a "dynamic" decreasing rate (we called "dynamic" because the value is recalculated during every iteration) where the level will be decreased based on different values (note that the decreasing rate in a standard great deluge is a predefined constant). Also, the success of the approach is due to the ability of the algorithm in exploring different region of the solution space i.e. our algorithm works on 50 different solutions at every iteration.

6   Conclusion and Future Work

This paper presents a hybridization approach that combined an electromagnetism-like mechanism with great deluge algorithm. To our knowledge, this is the first such algorithm aimed at this problem domain. In order to test the performance of our approach, experiments are carried out based on examination timetabling problems and compare with a set of state-of-the-art methods from the literature. In future work, efforts will be made to establish, compare and report on timings in relation to previously reported literature.

This approach is simple yet effective, and produced a number of best known results in comparison with other approaches studied in the literature. With the help of the dynamic decreasing rate that works on a set of different solutions,we believe our approach is capable in finding better solutions for the examination timetabling problem. With the increasing complexity of examination timetabling problems in many educational institutions, we believed that the proposed effective hybrid approach can be adapted with new constraints easily. Thus the application of this technique to ITC2007 and other recently formulations of the problem will be the subject of future work.

References

1. Abdullah, S., Burke, E.K.: A Multi-start large neighbourhood search approach with local search methods for examination timetabling. In: International Conference on Automated Planning and Scheduling (ICAPS 2006), Cumbria, UK, pp. 334–337 (2006)
2. Abdullah, S., Ahmadi, S., Burke, E.K., Dror, M.: Investigating Ahuja-Orlin's large neighbourhood search approach for examination timetabling. OR Spectrum 29(2), 351–372 (2007)
3. Birbil, S.I., Fang, S.C.: An electromagnetism-like mechanism for global optimization. Journal of Global Optimization 25, 263–282 (2003)

4. Burke, E.K., Elliman, D.G., Ford, P.H., Weare, R.F.: Examination timetabling in British universities - A survey. In: Burke, E.K., Ross, P. (eds.) PATAT 1995. LNCS, vol. 1153, pp. 76–92. Springer, Heidelberg (1996)
5. Burke, E.K., Newall, J.P., Weare, R.F.: A memetic algorithm for university exam timetabling. In: Burke, E.K., Ross, P. (eds.) PATAT 1995. LNCS, vol. 1153, pp. 241–250. Springer, Heidelberg (1996)
6. Burke, E.K., Bykov, Y., Newall, J.P., Petrovic, S.: A time-predefined local search approach to exam timetabling problem. IIE Transactions 36(6), 509–528 (2004)
7. Burke, E.K., Dror, M., Petrovic, S., Qu, R.: Hybrid graph heuristics in hyper-heuristic applied to exam timetabling problems. In: Golden, B.L., Raghavan, S., Wasil, E.A. (eds.) The next wave in computing, optimization and decision technologies, pp. 79–91. Springer, Maryland (2005)
8. Burke, E.K., Eckersley, A.J., McCollum, B., Petrovic, S., Qu, R.: Hybrid variable neighbourhood approaches to university exam timetabling. Technical Report NOTTCS-TR-2006-2, School of Computer Science and Information Technology, University of Nottingham, United Kingdom (2006)
9. Caramia, M., Dell'Olmo, P., Italiano, G.F.: New algorithms for examination timetabling. In: Näher, S., Wagner, D. (eds.) WAE 2000. LNCS, vol. 1982, pp. 230–241. Springer, Heidelberg (2001)
10. Carter, M.W., Laporte, G., Chinneck, J.W.: A general examination scheduling system. Interfaces 24(3), 109–120 (1994)
11. Carter, M.W., Laporte, G.: Recent developments in practical examination timetabling. In: Burke, E.K., Ross, P. (eds.) PATAT 1995. LNCS, vol. 1153, pp. 3–21. Springer, Heidelberg (1996)
12. Carter, M.W., Laporte, G., Lee, S.: Examination timetabling: Algorithmic strategies and applications. Journal of the Operational Research Society 47(3), 373–383 (1996)
13. Casey, S., Thompson, J.: GRASPing the examination scheduling problem. In: Burke, E.K., De Causmaecker, P. (eds.) PATAT 2002. LNCS, vol. 2740, pp. 232–244. Springer, Heidelberg (2003)
14. Côté, P., Wong, T., Sabourin, R.: A hybrid multi-objective evolutionary algorithm for the uncapacitated exam proximity problem. In: Burke, E.K., Trick, M.A. (eds.) PATAT 2004. LNCS, vol. 3616, pp. 294–312. Springer, Heidelberg (2005)
15. Dueck, G.: New optimisation heuristics: The great deluge algorithm and the record-to-record travel. Journal of Computational Physics 104, 86–92 (1993)
16. Lewis, R.: A survey of metaheuristic-based techniques for university timetabling problems. OR Spectrum 30(1), 167–190 (2008)
17. McCollum, B., Schaerf, A., Paechter, B., McMullan, P., Lewis, R., Parkes, A., Di Gaspero, L., Qu, R., Burke, E.K.: Setting the research agenda in automated timetabling: The second international timetabling competition. Accepted for publication to INFORMS Journal of Computing (2009), doi:10.1287/ijoc.1090.0320
18. McCollum, B., McMullan, P., Burke, E.K., Parkes, A.J., Qu, R.: A New Model for Automated Examination Timetabling. Accepted to Annals of OR. Post Proceedings of PATAT 2007, Montreal (2007)
19. Merlot, L.T.G., Boland, N., Hughes, B.D., Stuckey, P.J.: A hybrid algorithm for the examination timetabling problem. In: Burke, E.K., De Causmaecker, P. (eds.) PATAT 2002. LNCS, vol. 2740, pp. 207–231. Springer, Heidelberg (2003)
20. Petrovic, S., Burke, E.K.: University timetabling. In: Leung, J. (ed.) Handbook of Scheduling: Algorithms, Models and Performance Analysis, ch. 45. CRC Press, Boca Raton (2004)

21. Qu, R., Burke, E.K.: IHybridisations within a Graph Based Hyper-heuristic Framework for University Timetabling Problems. To appear at Journal of Operational Research Society, JORS (2008), doi:10.1057/jors.2008.102
22. Qu, R., Burke, E.K., McCollum, B.: Adaptive Automated Construction of Hybrid Heuristics for Exam Timetabling and Graph Colouring Problems. European Journal of Operational Research (EJOR) 198(2), 392–404 (2009)
23. Yang, Y., Petrovic, S.: A novel similarity measure for heuristic selection in examination timetabling. In: Burke, E.K., Trick, M.A. (eds.) PATAT 2004. LNCS, vol. 3616, pp. 247–269. Springer, Heidelberg (2005)

Iterative Relaxation-Based Heuristics for the Multiple-choice Multidimensional Knapsack Problem

Saïd Hanafi[1,2,3], Raïd Mansi[1,2,3], and Christophe Wilbaut[1,2,3]

[1] Univ Lille Nord de France, F-59000 Lille, France
[2] UVHC, LAMIH, F-59313 Valenciennes, France
[3] CNRS, UMR 8530, F-59313 Valenciennes, France
{said.hanafi,raid.mansi,christophe.wilbaut}@univ-valenciennes.fr

Abstract. The development of efficient hybrid methods for solving hard optimization problems is not new in the operational research community. Some of these methods are based on the complete exploration of small neighbourhoods. In this paper, we apply iterative relaxation-based heuristics that solves a series of small sub-problems generated by exploiting information obtained from a series of relaxations to the multiple–choice multidimensional knapsack problem. We also apply local search methods to improve the solutions generated by these algorithms. The method is evaluated on a set of problem instances from the literature, and compared to the results reached by both Cplex solver and an efficient column generation–based algorithm. The results of the method are encouraging with 9 new best lower bounds among 33 problem instances.

1 Introduction

Some recent and efficient hybrid methods can be found in the literature. Some of these approaches are based on an old idea in optimization which consists of generating solutions by iteratively holding selected subsets of variables fixed at particular values while varying the values of other variables.

For instance the Local Branching (**LB**) framework [6] and the Relaxation Induced Neighbourhood Search [4], proposed respectively by Fischetti and Lodi and by Danna et al. are based on the exploration of a small-size neighbourhood generated at each node of a branch-and-bound (or a branch-and-cut) search tree from the incumbent feasible solution. An important component of this kind of approach is the generation of the promising neighbourhoods at each step of the algorithm. Hansen et al. [7] combined the LB approach with the Variable Neighbourhood Search one to solve the mixed integer programming problem (**MIP**). Other efficient hybrid methods can be referred as the global intensification scheme proposed in [21]. This approach combines Tabu Search (**TS**) with Dynamic Programming (**DP**). The forward phase of DP is used to construct a list L by solving exactly a family of small sub problems. Then a second level of

M.J. Blesa et al. (Eds.): HM 2009, LNCS 5818, pp. 73–83, 2009.
© Springer-Verlag Berlin Heidelberg 2009

intensification is introduced in TS by implicitly solving a small reduced problem to evaluate the neighbourhood of the current solution (by scanning the list L in the backtracking phase of DP). In this paper we explore the use of iterative relaxation-based heuristics for the multiple–choice multidimensional knapsack problem. As we show in the continuation these algorithms provide both upper bounds and lower bounds of the input problem.

The rest of the paper is organized as follows. In Section 2 we present the main principles of the algorithms we propose. Section 3 describes the application of these algorithms to the multiple–choice multidimensional knapsack problem. The computational results are presented and discussed in Section 4 in an effort to assess the performance of the proposed methods. The last section is devoted to our conclusions.

2   Main Principles of the Iterative Heuristics

Wilbaut and Hanafi [20] proposed recently new convergent iterative heuristics for 0–1 MIP. The 0–1 MIP problem P can be expressed as

$$(P) \begin{bmatrix} \max cx = \sum_{j=1}^{n'} c_j x_j \\ \text{subject to:} \sum_{j=1}^{n'} a_{ij} x_j \le b_i \ \forall i \in M = \{1, \dots, m\} \\ x_j \in \{0, 1\} \quad j \in N = \{1, \dots, n\} \\ x_j \ge 0 \qquad j = n+1, \dots, n' \end{bmatrix}$$

where N is the set of binary variables and $n = |N|$. We also assume that the problem is feasible and that all constant c_j, a_{ij} and b_i are nonnegative.

For convenience, we first recall the main principles of the Iterative Heuristics (**IH**) described in [20]. The general method, which is a relaxation-based one, can be summarized by the following scheme in three steps:

- Step 1: Solve one or more relaxation(s) of the current problem and keep one or more optimal solution(s). Update the upper bound of the problem.
- Step 2: Solve one or more reduced problem induced by the previous solution(s), and eventually update the lower bound of the problem.
- Step 3: If a stopping condition is satisfied then return the best bounds of the problem. Otherwise add one or more pseudo-cuts generated from the solution(s) of the relaxation(s) and go to Step 1.

In the implementation of such a method we have to define the relaxation(s) used in Step 1 and the notion of reduced problem from Step 2. In this paper we deal with three variants based on the Linear Programming (**LP**) relaxation and/or a MIP-relaxation. To simplify the presentation of the approach we suppose that we use the LP-relaxation. We call the associated algorithm the Iterative Linear Programming-based Heuristic (**ILPH**).

A reduced problem as mentioned in Step 2 is defined from the original problem by setting some variables at given values. Formally, given a vector $y \in [0,1]^n$, let $J^0(y) = \{j \in N : y_j = 0\}$, $J^1(y) = \{j \in N : y_j = 1\}$, $J^*(y) = \{j \in N : y_j \in]0,1[\}$ and $J(y) = \{j \in N : y_j \in \{0,1\}\}$. Let P be an optimisation problem and C be a set of constraints; $(P|C)$ denotes the optimisation problem obtained from P by adding the constraint set C to P. The reduced problem associated to a vector y can be defined as follows

$$P(y) = (P|x_j = y_j, \forall j \in J(y)). \tag{1}$$

The ILPH restricts the search process to visiting optimal LP-solutions already generated by adding a pseudo-cut at each iteration according to the following propositions (the proofs can be found in [8]).

Proposition 1. *Let y be a vector in $\{0,1\}^n$ of a 0–1 integer problem. The following inequality*

$$\sum_{j \in J^1(y)} x_j - \sum_{j \in J^0(y)} x_j \leq |J^1(y)| - 1 \tag{2}$$

cuts off solution y without cutting off any other solution in $\{0,1\}^n$.

The next proposition is an extension of Proposition 1 for 0–1 MIP.

Proposition 2. *Given a 0-1 MIP problem P, let $\bar{x}$ be an optimal solution of the LP-relaxation $LP(P)$ and y be an optimal solution for the reduced problem $P(\bar{x})$. Thus, an optimal solution for P is either the feasible solution y or an optimal solution for the following problem*

$$\left(P\left|\left\{ \sum_{j \in J^1(\bar{x})} x_j - \sum_{j \in J^0(\bar{x})} x_j \leq |J^1(\bar{x})| - 1 \right\}\right.\right) \tag{3}$$

Proposition 2 is a basis of the proof for the finite convergence of the ILPH ([8,20]).

We give an algorithmic description of the ILPH in Algorithm 1. At each iteration, the ILPH algorithm attemps to improve the lower and upper bound on the optimal value of the given problem P. In this algorithm, the upper bound corresponds to the optimal value for the LP-relaxation of the current problem Q, and the lower bound corresponds to the value of the best known solution x^*. The improvements of the upper bound are assured by adding pseudo cuts (3) to the problem Q. The improvements of the lower bound are attempted by solving the reduced problem $P(\bar{x})$ induced from the LP-solution $\bar{x}$. The process is repeated until the stopping condition is satisfied. Then it returns the best bounds of the problem. Generally we can run the algorithm with a maximum number of iterations Max_{Iter} or stop it when the gap between the upper and lower bounds is less than a tolerance ϵ. The finite convergence of ILPH can be assured under non restrictive conditions (see [8,20] for more details).

Algorithm 1. The ILPH

Input: A problem P
$Q \leftarrow P$ current problem();
$x^* \leftarrow$ an initial feasible solution ;
$iter \leftarrow 0$, $Stop \leftarrow False$;
while $Not\ Stop$ **do**
 $iter \leftarrow iter + 1$;
 $\bar{x} \leftarrow$ an optimal solution of LP(Q) ;
 $x \leftarrow$ an optimal solution of $P(\bar{x})$;
 if $cx > cx^*$ **then**
  | $x^* \leftarrow x$;
 end

 $Q \leftarrow (Q \left| \{ \sum\limits_{j \in J^1(\bar{x})} x_j - \sum\limits_{j \in J^0(\bar{x})} x_j \leq \left| J^1(\bar{x}) \right| - 1 \} \right.)$;
 if $c\bar{x} - cx^* < \epsilon$ $or\ iter > Max_{iter}$ **then**
  | $Stop = True$
 end
end
return the best lower bound x^* and the best upper bound $\bar{x}$;

Wilbaut and Hanafi proposed in [20] several variants of the IH. They introduced a MIP-relaxation defined for the problem P relatively to a subset J of N as:

$$MIP(P, J) \left[\begin{array}{l} \max \sum\limits_{j=1}^{n'} c_j x_j \\ \text{s.t.: } \sum\limits_{j=1}^{n'} a_{ij} x_j \leq b_i\ \forall i \in M \\ \quad x_j \in [0, 1] \quad\quad j \in N \\ \quad x_j \geq 0 \quad\quad\quad j = n+1, \ldots, n' \\ \quad x_j \in \{0, 1\} \quad\quad j \in J \end{array} \right.$$

From this relaxation it is possible to define several heuristics in addition to the ILPH one. In particular we work in this paper with the following variants: the Iterative MIP-based Heuristic (**IMIPH**) in which the MIP-relaxation simply replaces the LP-relaxation; the Iterative Relaxations-based Heuristic (**IRH**) and the Independent Iterative Relaxations-based Heuristic (**IIRH**). In the IRH at a given iteration we first solve the LP-relaxation of the current problem P. Then we solve the MIP-relaxation defined by $MIP(P, J^*(\bar{x}))$, where $\bar{x}$ denotes an optimal solution of the LP-relaxation. These two relaxations generate two reduced problems and two pseudo-cuts. The reduced problems are solved exactly (or heuristically according to the chosen strategy), and the pseudo-cuts are added to enriche problem P. In the IIRH we define two problems P and P' associated respectively to the LP-relaxation and to the MIP-relaxation. These two problems are managed independently in the sense that we use the LP-relaxation only for problem P and we use the MIP-relaxation only for problem P'. This process can be viewed as a parallel version based simultaneously on the ILPH and on the IMIPH (in practice the two processes are managed in a sequential order).

In the next section we present the multiple–choice multidimensional knapsack problem, which is used as a platform in this paper.

3   The Multiple-choice Multidimensional Knapsack Problem

In this paper we consider the 0–1 Multiple–choice Multidimensional Knapsack Problem (**MMKP**), which seeks to find a subset of items that maximizes a linear objective function while satisfying a set of capacity constraints and a set of choice constraints. More precisely, in the MMKP we have a multi-constrained knapsack of a capacity vector $b = (b^1, b^2, \ldots, b^m)$, and a set $N = (N_1, N_2, \ldots, N_n)$ of n disjoint classes, where each class k has $n_k = |N_k|$ items for $k = 1, \ldots, n$. Each item j of class k has a nonnegative profit value c_{kj} for $j = 1, \ldots, n_k$, and requires nonnegative resources given by the weight vector $A = (a_{kj}^1, a_{kj}^2, \ldots, a_{kj}^m)$. The MMKP can be formulated as follows:

$$
(\text{MMKP}) \begin{cases} \max \sum_{k=1}^{n} \sum_{j \in N_k} c_{kj} x_{kj} \\ \text{s.t.} : \sum_{k=1}^{n} \sum_{j \in N_k} a_{kj}^i x_{kj} \leq b^i \ \forall i = 1, \ldots, m \\ \sum_{j \in N_k} x_{kj} = 1 \qquad k = 1, \ldots, n \\ x_{kj} \in \{0,1\} \qquad k = 1, \ldots, n, \ j \in N_k \end{cases}
$$

where x_{kj} equals 1 (respectively 0) when item j of the k-th class is picked (resp. not picked). The total number of variables is then given by $S_n = \sum_{k=1}^{n} n_k$.

Note that the MMKP is a variant of the well-known Multidimensional Knapsack Problem (**MKP**) [22]. The MMKP is also a generalization of the multiple–choice knapsack problem (**MCKP**) in which we have only one resource contraint.

Some applications of the MMKP can be found in the literature. For instance it was used in real-time decision context such as in quality adaptation and admission control of interactive multimedia systems [2,10] or in service level agreement management in telecommunication networks [9,19]. Several variants of the MCKP and the MMKP have also been proposed such as the budgeting problem with bounded multiple-choice constraints which was modeled by Pisinger [16] as a generalization of the MCKP. Li et al. [11] modeled the strike force asset allocation problem as a problem containing multiple knapsack constraints and multiple choice constraints.

A few papers that deal with both the MCKP and MMKP are available in the literature. For the MCKP some exact methods can be listed as in [17] with a branch-and-bound method, or in [5] where the authors proposed a hybrid dynamic programming - branch and bound algorithm. Pisinger also proposed in [15] an efficient method based on the solving of a core problem with a dynamic

programming method. One of the first reference for the MMKP is the heuristic
based on Lagrange multipliers due to Moser et al. [12]. It is also possible to find
papers in which the authors present the application of a method already proposed
for solving the MKP. This is the case for instance in [10] in which the authors
applied an algorithm based on the aggregate resources already introduced by
Toyoda [18], or in [13] where the authors extended the well-known approach of
Pirkul [14]. More recently, Hifi et al. [9] proposed a guided local search in which
the trajectories of the solutions were oriented by increasing the cost function
with a penalty term. Akbar et al. [1] reduced the multidimensional resource
constraints of the MMKP to a single constraint by using a penalty vector. Finally,
Cherfi and Hifi [3] proposed an efficient method based on a column generation
approach combined with a heuristic search procedure for solving the MMKP.
They improved some of the best-known values on a set of problem instances of
the literature, and we will use this algorithm as a reference in our numerical
experiments.

4   Computational Results

We give in this section the results obtained by applying several versions of the
IH described in the previous section: ILPH, IRH and IIRH. We do not present
the results of the IMIPH since preliminary experiments showed that this version
is clearly dominated by the others. The aim is to evaluate the performance of
our method compared to the results reached by the best version of the algorithm
proposed in [3].

Our algorithms were coded in C++ and all considered algorithms were tested
on a Pentium IV (3.4GHz and with 4 Gb of RAM). We used CPLEX of Ilog
as MIP-solver. To evaluate our algorithms we use the benchmarks mentioned in
the recent papers [3,9] and available on the Internet. The characteristics of these
instances are summarized in Tables 1 and 2. We tested a total of 33 instances
corresponding to two groups. In the first one we have 13 instances (noted I01,...,
I13) varying from small to large-scale size ones. In the second one we have 20
problem instances (noted Ins01, ..., Ins20) varying from medium to large size
ones. We report in Tables 1 and 2 the number of classes (resp. the number of

Table 1. First group of instances

Instance	n	n_k	m	S_n	Instance	n	n_k	m	S_n
I01	5	5	5	25	I08	10	150	10	1500
I02	5	10	5	50	I09	10	200	10	2000
I03	10	15	10	150	I10	10	250	10	2500
I04	10	20	10	200	I11	10	300	10	3000
I05	10	25	10	250	I12	10	350	10	3500
I06	10	30	10	300	I13	10	400	10	4000
I07	10	100	10	1000					

Table 2. Second group of instances

Instance	n	n_k	m	S_n	Instance	n	n_k	m	S_n
Ins01	50	10	10	500	Ins11	90	10	10	900
Ins02	50	10	10	500	Ins12	100	10	10	1000
Ins03	60	10	10	600	Ins13	100	30	10	3000
Ins04	70	10	10	700	Ins14	150	30	10	4500
Ins05	75	10	10	750	Ins15	180	30	10	5400
Ins06	75	10	10	750	Ins16	200	30	10	6000
Ins07	80	10	10	800	Ins17	250	30	10	7500
Ins08	80	10	10	800	Ins18	280	20	10	5600
Ins09	80	10	10	800	Ins19	300	20	10	6000
Ins10	90	10	10	900	Ins20	350	20	10	7000

Table 3. Illustration of the convergence of the ILPH

$iter$	ub	lb	$iter$	ub	lb
1	4812.82	**4799.3**	10	4804.07	4799.3
2	4811.04	4793.2	11	4803.81	4799.3
3	4810.6	4798.8	12	4803.12	4799.3
4	4807.26	4799.3	13	4802.59	4799.3
5	4805.93	4799.3	14	4801.67	4799.3
6	4805.64	4795.4	15	4801.54	4799.3
7	4805.3	4799.3	16	4800.97	4799.3
8	4805.08	4799.3	17	**4799.47**	4798.1
9	4804.42	4799.3			

items in each class) in column n (resp. n_k), the number of resource constraints in column m and the total number of variables in column S_n.

To illustrate the progress of the IH for the MMKP, we first report in Table 3 a trace of the execution of the ILPH on the medium size instance I06 with 300 variables (30 classes with 10 items by class). In Table 3 we give for several iterations (column $iter$) the value of the linear programming relaxation of the current problem in column ub, and the value of the solution obtained when solving the reduced problem in column lb.

For this instance, we can observe that the ILPH visits an optimal solution at the first iteration with a value equal to 4799.3. However the method needs to perform 17 iterations to prove the optimality of this solution. As we said in Section 2 it is a drawback of the ILPH if it is used as an exact method. Nevertheless the overall results given in this section show that this algorithm is able to visit strong lower bounds quickly for the MMKP.

We report in Table 4 the overall results obtained by our algorithms over the 33 instances. The aim of these experiments is first to show that the methods are able to visit good feasible solutions in a short time. Second we want to see if the algorithms can obtain optimal or near optimal solutions if we increase the CPU time.

Let us note that we also apply a simple local search method when the current best solution is improved during our algorithm. The local search consists in applying a local swapping strategy (the so-called 2-opt strategy) between two items j and j' belonging to the same class N_k.

We give in Table 4 for each instance the best-known value (or the optimal value when it is known) reported in [3] in column C&H. We then provide the values obtained by the ILPH, the IRH and the IIRH respectively.

Table 4. Overall results on the 33 instances

Instance	C&H	$cond_1$			$cond_2$		
		ILPH	IRH	IIRH	ILPH	IRH	IIRH
I1	173	173*	173*	173*	173*	173*	173*
I2	364	364*	364*	364*	364*	364*	364*
I3	1602	*1602*	*1602*	*1602*	*1602*	*1602*	*1602*
I4	3597	*3597*	*3597*	*3597*	*3597*	*3597*	*3597*
I5	3905,7	3905,7*	3905,7*	3905,7*	3905,7*	3905,7*	3905,7*
I6	4799,3	4799,3*	4799,3*	4799,3*	4799,3*	4799,3*	4799,3*
I7	**24587**	24585	24586	24585	24585	24586	24585
I8	**36892**	36866	36875	36875	36866	36883	36883
I9	**49176**	49148	49165	49172	49148	49172	49172
I10	61461	61442	61450	61446	61442	61453	**61465**
I11	**73775**	73761	73767	73766	73761	73770	73769
I12	**86078**	86065	86073	86061	86077	86073	86072
I13	**98431**	98407	98410	98428	98421	98414	98428
Avg. Gap (%)		0.017	0.01	0.009	0.015	0.006	0.004
Inst1	**10732**	10712	10702	10714	10712	10714	10714
Inst2	**13598**	13596	13597	13597	13596	13597	13597
Inst3	10943	10934	10934	10934	10934	*10943*	10934
Inst4	14440	**14442**	14434	14430	**14442**	14439	14441
Inst5	17053	17041	17042	17037	17041	17050	**17061**
Inst6	**16825**	16823	16823	16820	16823	16823	16823
Inst7	**16435**	16408	16417	16434	16420	16417	16434
Inst8	**17510**	17503	17497	17489	17503	17497	17503
Inst9	**17760**	17744	17743	17751	17744	17753	17751
Inst10	**19314**	19298	19295	19295	19298	19306	19311
Inst11	**19434**	19418	19427	19419	19418	19430	19426
Inst12	21731	21720	**21738**	21724	21720	**21738**	**21738**
Inst13	**21575**	21573	21573	21570	21574	21574	21572
Inst14	**32870**	32869	32868	32868	32869	32869	32868
Inst15	39157	39155	39155	39155	**39160**	39157	39157
Inst16	43361	**43363**	43355	43361	**43363**	43360	43361
Inst17	54349	54351	*54352*	54351	54351	54354	**54356**
Inst18	60460	60458	60456	*60460*	**60462**	60461	60460
Inst19	64923	*64924*	64923	64918	64924	64923	**64925**
Inst20	**75611**	75606	75609	75604	75606	75609	75604
Avg. Gap (%)		0.044	0.045	0.044	0.039	0.023	0.017

According to the previous description of our objectives we report the results for two kind of tests: in the first one (corresponding to $cond_1$ in Table 4) we fix the allowed CPU time to 2 minutes. In the second one ($cond_2$) we use a different stopping condition. The algorithms stop when the CPU time comes to 5 minutes or when 100 iterations are performed. In Table 4 a bold value means that the corresponding algorithm provides the best-known value, and a * means that the algorithm proved the optimality of the corresponding solution. Finally rows "Avg. Gap" provide the average gap of our lower bounds with those reported in [3]. In this paper the authors tested several variants of their algorithm on a UltraSparc 10 (250Mhz with 1Gb of RAM). The CPU times varied from less than 1 sec. to 6 minutes for the first data set. They varied from 15 sec. to 12 minutes for the second data set.

Some conclusions can be listed from the results presented in Table 4:

- Our three algorithms are able to prove the optimality of the visited solution for 4 (small) instances (in a few seconds).
- Even if our algorithms can not be used in practice as exact method (in a reasonable CPU time), they can produce very interesting results in a short CPU time.
- Generally the results show that our methods are more efficient for the second group of instances, and for the large instances (in particular if we do not consider the first instances in the first group).

The left part of the table shows that our algorithms provide 5 new best-known values with a maximum CPU time of 2 minutes. They also visit 7 best solutions. That leads to confirm that these methods are efficient to provide good lower bounds of the MMKP very quickly. The average gap with the algorithm of Cherfi and Hifi is around 0.01% for data set 1 and around 0.04% for data set 2.

The right part of the table shows that if we increase the total CPU time of the IH then they provide some other new best-known values. This is in particular the case for the larger instances. Let us note that in this case the IRH seems to be less efficient. On the contrary the IIRH provides 5 new best values, and the ILPH provides 3 new best values. The first group of instances seems to be more difficult for our algorithms.

To summarize the results we can say that our algorithms visit 9 new best solutions, 4 best solutions and prove the optimality of 4 other solutions in a maximum CPU time of 5 minutes, which seems to be a reasonable CPU time for a heuristic when solving large instances. Our algorithms are able to improve the best values for several large instances, which is another interesting point.

5   Conclusions

In this paper we solved the multiple–choice multidimensional knapsack problem using iterative relaxations-based heuristics. These algorithms, which can be used theorically as exact methods, solve a series of small sub-problems generated by exploiting information obtained from a series of relaxations.

Computational results show that the process is able to provide interesting lower bounds quickly for all the instances. When we increase the maximum number of iterations the algorithm can also provide better solutions. In particular the methods provide 9 new best-known values over 33 instances taken from the literature.

We have some ideas to try to improve this work. Experiments showed that the ILPH converges very quickly to good lower bounds. On the contrary the other variants improve their solutions when the number of iterations increases. Thus we think it could be possible to combine more efficiently the different relaxations in our algorithms. We also want to take more into account the characteristic of the MMKP, i.e. the choice constraints. Finally we want to explore the use of local search in a better way since classic local search methods seems to be insufficient to really improve the lower bounds produced by our algorithms.

Acknowledgments

The present research work has been supported by International Campus on Safety and Intermodality in Transportation, the Nord-Pas-de-Calais Region, the European Community, the Regional Delegation for Research and Technology, the Ministry of Higher Education and Research, and the National Center for Scientific Research. The authors gratefully acknowledge the support of these institutions.

The authors also thank the anonymous referees for their comments which contributed to the improvement of the presentation of the paper.

References

1. Akbar, M.M., Rahman, M.S., Kaykobad, M., Manning, E.G., Shoja, G.C.: Solving the multidimensional multiple-choice knapsack problem by constructing convex hulls. Computers & Operations Research 33, 1259–1273 (2006)
2. Chen, L., Khan, S., Li, K.F., Manning, E.G.: Building an adaptive multimedia system using the utility model. In: Rolim, J.D.P. (ed.) IPPS-WS 1999 and SPDP-WS 1999. LNCS, vol. 1586, pp. 289–298. Springer, Heidelberg (1999)
3. Cherfi, N., Hifi, M.: A column generation method for the multiple–choice multi–dimensional knapsack problem. Computational Optimization and Applications (2008), doi:10.1007/s10589-008-9184-7
4. Danna, E., Rothberg, E., Le Pape, C.: Exploring relaxations induced neighborhoods to improve MIP solutions. Mathematical Programming 102, 71–90 (2005)
5. Dyer, M.E., Riha, W.O., Walker, J.: A hybrid dynamic programming/branch and bound algorithm for the multiple–choice knapsack problem. European Journal of Operational Research 58, 43–54 (1995)
6. Fischetti, M., Lodi, A.: Local Branching. Mathematical Programming 98, 23–47 (2003)
7. Hansen, P., Mladenovic, N., Urosevic, D.: Variable neighbourhood search and local branching. Computers & Operations Research 33, 3034–3045 (2006)

8. Hanafi, S., Wilbaut, C.: Improved convergent heuristics for the 0–1 multidimensional knapsack problem. Annals of Operations Research (2009), doi:10.1007/s10479-009-0546-z
9. Hifi, M., Michrafy, M., Sbihi, A.: Heuristic algorithms for the multiple–choice multidimensional knapsack problem. Journal of the Operational Research Society 55, 1323–1332 (2004)
10. Khan, S., Li, K.F., Manning, E.G., Akbar, M.M.: Solving the knapsack problem for adaptive multimedia systems. Studia Informatica Universalis 2, 157–178 (2002)
11. Li, V.C., Curry, G.L., Boyd, E.A.: Towards the real time solution of strike force asset allocation problems. Computers & Operations Research 31, 273–291 (2004)
12. Moser, M., Jokanovic, D.P., Shiratori, N.: An algorithm for the multidimensional multiple–choice knapsack problem. IEICE Transactions A(E80), 582–589 (1997)
13. Parra-Hernandez, R., Dimopoulos, N.: A new heuristic for solving the multi-choice multidimensional knapsack problem. Technical Report, Department of Electrical and Computer Engineering, University of Victoria (2002)
14. Pirkul, H.: A heuristic solution procedure for the multiconstraint zero–one knapsack problem. Naval Research Logistics 34, 161–172 (1987)
15. Pisinger, D.: A minimal algorithm for the multiple–choice knapsack problem. European Journal of Operational Research 83, 394–410 (1995)
16. Pisinger, D.: Budgeting with bounded multiple-choice constraints. European Journal of Operational Research 129, 471–480 (2001)
17. Sinha, A., Zoltners, A.: The multiple–choice knapsack problem. Operations Research 27, 503–515 (1979)
18. Toyoda, Y.: A simplified algorithm for obtaining approximate solution to zero–one programming problems. Management Sciences 21, 1417–1427 (1975)
19. Watson, R.K.: Packet Networks and optimal admission and upgrade of service level agreements: applying the utility model. M.A.Sc. Thesis, Department of ECE, University of Victoria (2001)
20. Wilbaut, C., Hanafi, S.: New convergent heuristics for 0–1 mixed integer programming. European Journal of Operational Research 195, 62–74 (2009)
21. Wilbaut, C., Hanafi, S., Fréville, A., Balev, S.: Tabu Search: Global Intensification using Dynamic Programming. Control and Cybernetics 35, 579–598 (2006)
22. Wilbaut, C., Hanafi, S., Salhi, S.: A survey of effective heuristics and their application to a variety of knapsack problems. IMA Journal of Management Mathematics 19, 227–244 (2008)

Solving a Video-Server Load Re-Balancing Problem by Mixed Integer Programming and Hybrid Variable Neighborhood Search*

Jakob Walla, Mario Ruthmair, and Günther R. Raidl

Institute of Computer Graphics and Algorithms,
Vienna University of Technology, Vienna, Austria
walla@nosystem.net, {raidl,ruthmair}@ads.tuwien.ac.at

Abstract. A Video-on-Demand system usually consists of a large number of independent video servers. In order to utilize network resources as efficiently as possible the overall network load should be balanced among the available servers. We consider a problem formulation based on an estimation of the expected number of requests per movie during the period of highest user interest. Apart from load balancing our formulation also deals with the minimization of reorganization costs associated with a newly obtained solution. We present two approaches to solve this problem: an exact formulation as a mixed-integer linear program (MIP) and a metaheuristic hybrid based on variable neighborhood search (VNS). Among others the VNS features two special large neighborhood structures searched using the MIP approach and by efficiently calculating cyclic exchanges, respectively. While the MIP approach alone is only able to obtain good solutions for instances involving few servers, the hybrid VNS performs well especially also on larger instances.

1 Introduction

Over the last few years internet-based video-on-demand (VoD) services have become increasingly popular. In contrast to traditional web- and file-services, a VoD service must reserve a certain amount of bandwidth for each request in order to guarantee uninterrupted playback. Therefore operators of VoD services are faced with high costs for high-bandwidth network connections and server hardware. Hence existing bandwidth resources should be utilized as efficiently as possible in order to avoid acquisition of excess bandwidth and reduce costs.

Recent works in this field have mainly focused on distributed video server architectures. A distributed VoD system consists of multiple video servers, each server having a dedicated network link as well as a dedicated storage subsystem. Because of storage capacity constraints each server can only hold replicas of a subset of all available video files. On arrival of a user request a central dispatcher component selects a server holding a replica of the desired video file

---

* This work is supported by the Austrian Science Fund (FWF) under contract number P20342-N133.

M.J. Blesa et al. (Eds.): HM 2009, LNCS 5818, pp. 84–99, 2009.

with enough available bandwidth to handle the request. If no such server is available, the request must be rejected. Thus, a common design goal of VoD systems is to minimize the probability that a user request has to be rejected [1,2,3,4]. Zhou et al. [5] try to achieve this goal by maximizing the replication degree while at the same time minimizing the load imbalance degree.

In this work we present an approach to VoD load balancing based on a priori assignment of expected requests to servers. Besides minimization of load imbalance our formulation deals with a problem frequently encountered in real-world systems which to the best of the authors' knowledge has not yet been explicitly addressed in literature. After determining the assignments of predicted user requests and the according video-replica to the servers, the new replica assignment still has to be realized physically. This can lead to high amounts of data being transferred between the video servers causing considerable reorganization overhead as well as impairing system performance. Therefore our problem formulation aims at minimizing load imbalance while making a necessary reorganization phase as short as possible at the same time. We refer to this optimization problem as *Video-Server Load Re-Balancing* (VSLRB). More details on the approaches presented here can be found in the master thesis of the first author [6].

The next section defines the VSLRB problem more formally. Section 3 gives an overview on related work. A mixed integer programming formulation for solving small instances of the problem to proven optimality is introduced in Section 4. Our new hybrid variable neighborhood search approach for addressing larger instances is presented in Section 5. Section 6 discusses experimental results, and Section 7 concludes this article.

2   Problem Definition

We consider a VoD system consisting of a set C of m video servers hosting a set F of n video files. Furthermore, we are given a set T of video file types. Each video server $j \in C$ has associated a storage capacity $W_j > 0$ and upload and download transmission capacities of the server's network link denoted by $U_j > 0$ and $D_j > 0$, respectively. Finally each server j has a subset of video file types $T_j \subseteq T$ it accepts. In turn, each video file $i \in F$ has a certain file size $w_i > 0$, a bitrate $b_i > 0$ and a file type $t_i \in T$. Each server j holds a set of replicas $F_j \subseteq F$, where $t_i \in T_j$, $\forall i \in F_j$. Conversely each video file i is held by a set of servers $C_i \subseteq C$.

Some works specifically focus on modeling of user behavior in VoD systems [7,8,9]. A method for modeling video popularity combined with a method for modeling temporal distribution of user requests can be used to estimate the number of requests to file i during the daily peak period of user interest [5]. In this work we assume the availability of such an estimation and denote by $q_i \geq 0$, $\forall i \in F$, the estimated number of requests for video file i during the daily peak period. This allows for an estimation of the worst case load $L = \sum_{i \in F} q_i b_i$, i.e. when all requests predicted to occur during the peak period are active at the same time. The worst case load is to be balanced among the available servers by assigning the predicted requests.

This assignment of requests to servers is denoted by the assignment function $Q : F \times C \to \mathbb{N}_0$. Thus, $Q(i,j)$ denotes the amount of parallel requests for file i handled by server j. Any valid Q must satisfy the constraint $\sum_{j \in C_i} Q(i,j) = q_i$, $\forall i \in F$. Furthermore, Q must contain only valid assignments w.r.t. allowed file types; i.e. let $P = \{(i,j) \in F \times C \mid t_i \in T_j\}$, then

$$(i,j) \notin P \Rightarrow Q(i,j) = 0. \tag{1}$$

As a server j needs to hold a replica of a file i in order to handle requests for it, the concrete choice of Q determines the configuration of the sets of replicas:

$$Q(i,j) > 0 \Leftrightarrow i \in F_j \quad \text{and therefore} \quad Q(i,j) = 0 \Leftrightarrow i \notin F_j. \tag{2}$$

The *server load* $\mathcal{L}(j)$ is expressed as the total bandwidth requirement to fulfill all assigned requests:

$$\mathcal{L}(j) = \sum_{i \in F_j} b_i Q(i,j), \quad \forall j \in C. \tag{3}$$

The first goal of our assignment optimization is to minimize the sum of absolute deviations of server loads from given target loads:

$$\min \sum_{j \in C} |\eta_j - \mathcal{L}(j)|, \tag{4}$$

where the η_j are chosen in a way so that $\sum_{j \in C} \eta_j = L$ with respect to the accepted file types T_j. The target load values η_j are pre-calculated during the creation of an instance of VSLRB using a quadratic programming formulation, see [6] for details.

The second optimization goal is concerned with the minimization of the reorganization overhead imposed by a concrete assignment function Q. Let $\overline{F}_j$, $\forall j \in C$, denote the sets of replicas before applying the assignment optimization procedure. Whenever a replica of file i required by the newly obtained assignment of requests is not already present on a respective server j, i.e. $i \in F_j \wedge i \notin \overline{F}_j$, file i must be transferred to j causing undesirable excess network load. If such a transmission occurs it should be spread over as many source servers as possible in order to reduce network load on each of the source servers.

The time needed for the transmission of file i to server j can be estimated by

$$T(i,j) = \sum_{k \in \overline{C}_i} T(i,k,j), \tag{5}$$

where $\overline{C}_i$ denotes the set of servers currently holding file i and $T(i,k,j)$ the time needed to transfer the part of the file contributed by source server k. The size of the part server k contributes is proportional to its current share of the total load caused by file i:

$$T(i,k,j) = \frac{\overline{Q}(i,k)\, w_i}{\overline{q}_i\, c_{k,j}}, \tag{6}$$

where $\overline{Q}(i,k)$ denotes the number of requests for file i currently assigned to server k, $\overline{q}_i$ the overall number of requests considered for file i so far, and $c_{k,j} = \min\{U_k, D_j\}$ the possible transfer rate from server k to server j. Assuming all partial transmissions are carried out sequentially, the minimization of the duration of the reorganization phase can now be expressed as

$$\min \sum_{k \in C} \sum_{j \in C, j \neq k} \sum_{i \in (F_j \setminus \overline{F}_j) \cap \overline{F}_k} \mathcal{T}(i,k,j). \tag{7}$$

A valid assignment function Q must fulfill certain further restrictions. Firstly, no server is allowed to exceed its storage capacity, i.e.

$$\sum_{i \in F_j} w_i \leq W_j, \quad \forall j \in C. \tag{8}$$

Secondly, the inbound data volume of each server must be limited to the currently available storage capacity:

$$\sum_{i \in F_j \setminus \overline{F}_j} w_i \leq W_j - \sum_{i \in \overline{F}_j} w_i, \quad \forall j \in C. \tag{9}$$

Without this constraint a server might need to move outbound replicas before it can receive any inbound replicas potentially leading to a deadlock situation.

3 Related Work

Similarities exist between VSLRB and other VoD-specific optimization problems [5,10,11]. Other related problems arise in multi-processor scheduling [12,13].

Some works in literature employ a comparable formalization based on a static distribution of requests for replicas hosted on video servers but differ in their choice of the objective function. Chen et al. [10] focus on the bin-packing aspect of the problem, i.e. finding a minimal number of servers along with an assignment of replicas satisfying a given access profile. The authors describe an algorithm inspired by the transport simplex method for solving this problem. Wang et al. [11] describe a branch-and-bound algorithm as well as a greedy heuristic for a similar problem. Wolf et al. [4] describe a two-level procedure based on the theory of resource allocation problems. In a first step, a greedy heuristic is used to calculate a required number of replicas per video file. In a second step, these replicas are assigned to *Disk Striping Groups* (DSGs) so that the forecast load of any DSG is proportional to its stream capacity. Zhou et al. [5] focus on finding a load balanced solution for a fixed number of servers. Replicas are allowed to be recoded in order to reduce the bandwidth requirements of the according requests. The optimization goal is to find a replica assignment that maximizes the replication degree as well as the average bitrate and at the same time minimizes the load imbalance degree. For the special case of a single fixed bitrate the authors give an exact algorithm consisting of *bounded Adams' monotone divisor* replication and *smallest load first placement*. For the general case the authors propose a heuristic based on simulated annealing. Some parallels exist between VSLRB and special cases of the well-known multiprocessor scheduling problem. Aggerwal et al. [12] consider a variant called *load rebalancing problem*. Given a valid schedule along with job-specific relocation costs a new schedule with minimal makespan is to be obtained while the total relocation costs are constrained by a given bound. The authors describe an approximation algorithm as well as a polynomial-time approximation scheme.

Furthermore, in the terminology of a recent survey of scheduling problems by Allaverdi et al. [13] VSLRB can be considered as a sequence-independent batch multiprocessor scheduling problem. Requests to the same video file can be viewed as jobs of the same family while batch setup times correspond to the reorganization time necessary for placing a replica on a server. Despite this correspondence the authors do not mention an objective function comparable to the one of VSLRB.

4   Mixed Integer Programming Formulation

Given the formal definition of VSLRB from Section 2, we can model the problem as the following mixed integer linear program (MIP).

$$\min \quad \alpha \sum_{j \in C} y_j + \beta \sum_{k \in C} \sum_{j \in C,\, j \neq k} \frac{1}{c_{k,j}} \sum_{i \in F | t_i \in T_j} \frac{\overline{x}_k^i \, (1 - \overline{p}_j^i) \, w_i}{\overline{q}_i} \, p_j^i \tag{10}$$

subject to

$$\eta_j - \textstyle\sum_{i \in F | t_i \in T_j} b_i x_j^i \leq y_j, \qquad\qquad \forall j \in C \tag{11}$$

$$-\eta_j + \textstyle\sum_{i \in F | t_i \in T_j} b_i x_j^i \leq y_j, \qquad\qquad \forall j \in C \tag{12}$$

$$\textstyle\sum_{j \in C | t_i \in T_j} x_j^i = q_i, \qquad\qquad \forall i \in F \tag{13}$$

$$p_j^i - \tfrac{x_j^i}{q_i} \geq 0, \qquad\qquad \forall (i,j) \in P \tag{14}$$

$$p_j^i - \tfrac{x_j^i}{q_i} \leq 1 - \tfrac{1}{q_i}, \qquad\qquad \forall (i,j) \in P \tag{15}$$

$$\textstyle\sum_{i \in F | t_i \in T_j} w_i p_j^i \leq W_j, \qquad\qquad \forall j \in C \tag{16}$$

$$\textstyle\sum_{i \in F | t_i \in T_j} (1 - \overline{p}_j^i) \, w_i \, p_j^i \leq W_j - \textstyle\sum_{i \in F} \overline{p}_j^i w_i, \quad \forall j \in C \tag{17}$$

$$x_j^i \in \{0, \ldots, q_i\}, \qquad\qquad \forall (i,j) \in P \tag{18}$$

$$p_j^i \in \{0, 1\}, \qquad\qquad \forall (i,j) \in P \tag{19}$$

$$y_j \geq 0, \qquad\qquad \forall j \in C \tag{20}$$

The assignment function Q is expressed by non-negative integer decision variables $x_j^i = Q(i,j)$ and the sets of replicas by binary decision variables p_j^i, $\forall (i,j) \in P$, where $p_j^i = 1 \Leftrightarrow i \in F_j$. Corresponding constants $\overline{x}_j^i$ and $\overline{p}_j^i$

represent the previous state before the reassignment, respectively. The objective function (10) combines the two goal functions (4) and (7) in a linear fashion using weights $\alpha > 0$ and $\beta > 0$. Variables y_j together with inequalities (11) and (12) are used to model the absolute load deviations $|\eta_j - \mathcal{L}(j)|$, $\forall j \in C$, of (4). Constraints (14) and (15) define the relation between corresponding x_j^i and p_j^i variables expressed in the original problem formulation by (2). Eq. (14) enforces $p_j^i = 1$ if $x_j^i > 0$. Conversely, (15) enforces $x_j^i > 0$ if $p_j^i = 1$.

Set operations occurring in the original formulation in (7) and (9) are expressed in (10) and (17) by multiplying the respective decision variables with appropriate constants.

Eq. (13) ensures that no request for any $i \in F$ is left unassigned. Finally, (16) is used to model the storage capacity constraints expressed in the orginal formulation by (8).

Detailed experimental tests using ILOG CPLEX 11.1 for solving this MIP formulation clearly indicated that the performance substantially depends on the number of servers m, while the numbers of files and requests only have minor influence. In general, the approach yields good results in reasonable time only for a very small number of servers (less than 5), while the performance quickly deteriorates with larger m. For more details on these experiments we refer to [6]; selected results are also shown in Section 6.

5 Variable Neighborhood Search

Variable Neighborhood Descent (VND) [14] extends classical local search by systematically switching between multiple neighborhood structures $\mathcal{N}_1, \ldots, \mathcal{N}_{k_{\max}}$ in order to escape simple local optima and find better solutions that are optimal w.r.t. all these neighborhood structures. For an outline of the procedure see Alg. 5.1.

Variable Neighborhood Search (VNS) [14], shown in Alg. 5.2, is a metaheuristic that has a similar basic functionality but primarily addresses diversification and search space coverage. It also works on multiple neighborhood structures $N_1, \ldots, N_{l_{\max}}$ but these are typically larger than those of the VND and searched by just evaluating individual random moves; this process is called *shaking*. VNS contains an embedded local improvement procedure for intensification. This local improvement can be a simple local search or a more sophisticated procedure like VND. In the latter case, the VNS is called a *general VNS*.

In our specific general VNS for VSLRB, all of the employed neighborhood structures rely on the following two basic operations:

assign(i, j), $(i, j) \in P$: Assigns a request for video file $i \in F$ to server $j \in C$. If currently $Q(i, j) = 0$, i must be added to F_j.

unassign(i, j), $(i, j) \in P$: Unassigns a request for video file $i \in F$ from server $j \in C$. If $Q(i, j) = 0$ after the operation, i must be removed from F_j.

Algorithm 5.1. Variable Neighborhood Descent (VND)

Input: Initial solution x_s
$x \leftarrow x_s$
$l \leftarrow 1$
repeat
    | $x' \leftarrow$ search $\mathcal{N}_l(x)$ for a better or best neighbor
    | **if** $f(x') \leq f(x)$ **then**
    | | $x \leftarrow x'$
    | | $l \leftarrow 1$
    | **else**
    | $\lfloor$ $l \leftarrow l + 1$
until $l > l_{\max}$

Algorithm 5.2. Variable Neighborhood Search (VNS)

Input: Initial solution x_s
$x \leftarrow x_s$
$k \leftarrow 1$
repeat
    | **repeat**
    | | $x' \leftarrow$ pick random neighbor from $N_k(x)$ // *shaking*
    | | $x'' \leftarrow$ locally improve x'
    | | **if** $f(x'') \leq f(x)$ **then**
    | | | $x \leftarrow x''$
    | | | $k \leftarrow 1$
    | | **else**
    | | $\lfloor$ $k \leftarrow k + 1$
    | **until** $k > k_{\max}$
until Stopping criteria

In both cases, the objective function value is updated incrementally. In case of the second objective this can be achieved by pre-calculating the costs $\mathcal{R}(i, j)$ of placing a replica of i on server j:

$$\mathcal{R}(i, j) = \begin{cases} 0 & \text{if } i \in \overline{F}_j \\ \sum_{k \in \overline{C}_i} \mathcal{T}(i, k, j) & \text{otherwise} \end{cases} \tag{21}$$

The VND uses the following neighborhood structures in the listed order.

5.1 Access Move Neighborhood ($\mathcal{N}_{\mathrm{Move}}$)

The access move neighborhood contains all solutions Q reachable by moving a request for video file i assigned to some server j to another server k accepting type t_i. The operation **move**(i, j, k), $(i, j, k) \in \{F \times C \times C \mid t_i \in T_j \cap T_k\}$, therefore is defined by calling **unassign**(i, j) and **assign**(i, k). As there exist m possible source servers, at most n replicas on each source server, and at most $m - 1$ target servers, this neighborhood contains $\mathcal{O}(m^2 n)$ neighboring solutions.

5.2 Access Swap Neighborhood ($\mathcal{N}_{\text{Swap}}$)

This neighborhood contains all solutions Q reachable by swapping a request for video file i currently assigned to some server j with a request for a different file f currently assigned to a different server c. Thus, $\textbf{swap}(i, j, f, c)$, $(i, j, f, c) \in \{F \times C \times F \times C \mid t_i \in T_j \cap T_c \wedge t_f \in T_j \cap T_c\}$, performs the following basic operations: $\textbf{unassign}(i, j)$, $\textbf{unassign}(f, c)$, $\textbf{assign}(i, c)$ and $\textbf{assign}(f, j)$. When enumerating all possible neighboring solutions any assignment $(i, j) \in P$ needs to be considered only once for any two operations $\textbf{swap}(i, j, f, c)$ and $\textbf{swap}(f, c, i, j)$. As there are at most mn assignments and therefore no more than mn movable requests to consider, the size of the access swap neighborhood is bounded by $\frac{m(m-1)}{2} \frac{n(n-1)}{2} = \mathcal{O}(m^2 n^2)$.

5.3 κ-Server MIP Neighborhood ($\mathcal{N}_{\kappa-\text{MIP}}$)

This large neighborhood combines the VNS with the MIP approach described in Section 4. As already mentioned, the MIP approach in general only yields good results for instances involving a small number m of servers. Given an existing solution Q to an instance of VSLRB, we select a small number of κ servers in order to construct a subproblem that essentially is a smaller instance of VSLRB. Only the variables associated with these servers are to be optimized, all the others are fixed to their values of the current VNS solution and considered as constants. Let C' denote this set of selected servers. Then, the considered set of files and corresponding request amounts are

$$F' = \bigcup_{j \in C'} F_j, \quad \text{and} \quad q_i' = \sum_{j \in C'} Q(i, j), \quad \overline{q}_i' = \sum_{j \in C'} \overline{Q}(i, j), \quad \forall i \in F'. \quad (22)$$

The neighborhood of a current solution Q is implicitly defined as all feasible solutions to this subproblem. As κ is small, the MIP approach can be used to efficiently search this neighborhood.

A server selection C' leading to a promising subproblem must have two characteristics:

- C' has to include servers j with $\mathcal{L}(j) < \eta_j$ as well as servers $k \neq j$ with $\mathcal{L}(k) > \eta_k$.
- C' has to include at least two servers $j \neq k$ with $T_j \cap T_k \neq \emptyset$.

A subproblem without overlapping accepted file types is considered invalid because it does not allow for any improvement. For the task of selecting a set of servers C' we employ the greedy heuristic depicted in Alg. 5.3.

For any file $i \in F'$ there exist q_i' requests to be spread over at most $|A_i'|$ servers, where $A_i' = \{j \in C' \mid t_i \in T_j\}$ denotes the set of servers in the subproblem allowed to hold file i. As for each file i there exist

$$\binom{|A_i'| + q_i' - 1}{q_i'} \quad (23)$$

Algorithm 5.3. Select Servers

Input: A solution Q to an instance of VSLRB
sorted $\leftarrow$ sort servers $j \in C$ by descending $\mathcal{L}(j) - \eta_j$
$C' \leftarrow \emptyset$
coveredTypes $\leftarrow \emptyset$
for $l \leftarrow 1$ **to** $\lfloor \frac{\kappa}{2} \rfloor$ **do**
$\quad \mid \quad C' \leftarrow C' \cup sorted[l]$
$\quad \mid \quad coveredTypes \leftarrow T_{sorted[l]}$
$l \leftarrow m$
while $|C'| < \kappa \wedge l > \lfloor \frac{\kappa}{2} \rfloor$ **do**
$\quad \mid \quad$ **if** $coveredTypes \cap T_{sorted[l]} \neq \emptyset$ **then**
$\quad \mid \quad \quad \mid \quad C' \leftarrow C' \cup sorted[l]$
$\quad \mid \quad l \leftarrow l - 1$

---

possible assignment configurations, the size of the κ-server MIP neighborhood is bounded by

$$\prod_{i \in F'} \binom{|A'_i| + q'_i - 1}{q'_i} = \mathcal{O}\left(\prod_{i \in F} \binom{\kappa + q_i - 1}{q_i} \right). \tag{24}$$

5.4   Cyclic Exchange Neighborhood ($\mathcal{N}_{\text{Cyclic}}$)

A neighborhood structure based on cyclic exchanges of elements between subsets was first described by Thompson and Orlin [15]. Such a neighborhood structure can be applied to problems that can be naturally formulated as a partitioning problem.

Definition 1 (Generic Partitioning Problem). *We are given a finite set $A = \{a_1, a_2, \ldots, a_n\}$ of n elements and a cost function $c : \mathcal{P}(A) \to \mathbb{R}$, where $\mathcal{P}(A)$ denotes the power set of A. Furthermore we are given an integer $K \in \mathbb{N}^+$. Our goal is to find a K-partition $S = \{S_1, S_2, \ldots, S_K\}$ of mutually disjoint subsets S_i where $\bigcup_{i=1}^{K} S_i = S$, minimizing the total cost of $c(S) = \sum_{i=1}^{K} c(S_i)$. A total cost function that can be expressed in this way is said to be* separable over subsets.

Clearly, VSLRB can be formulated in such a way, with A corresponding to the entirety of all user requests and the subsets $S_1, \ldots, S_K$ corresponding to the servers $j \in C$. The cost $c(S_j)$ of a subset associated with server j can be calculated independently from the costs of the other servers by

$$c(S_j) = \alpha |\eta_j - \mathcal{L}(j)| + \beta \sum_{i \in F_j} \mathcal{R}(i, j) \tag{25}$$

A *cyclic exchange* or *cyclic transfer* is a simultaneous cyclic shift of up to K elements across up to K subsets. We adopt the notation of Ahuja et al. [16] to denote a cyclic exchange with $i_1 - i_2 - \ldots - i_r - i_1$. Each element $i_p \in A$, $p \in$

$\{1, \ldots, r\}$, is moved from $S[i_p]$ to $S[i_{p+1}]$, where $i_{r+1} = i_1$. We denote by $S[i_p]$ the subset which currently contains element i_p. The cost difference associated with inserting i_p in $S[i_{p+1}]$ and at the same time removing i_{p+1} from this set can be calculated by

$$c(S[i_{p+1}] \cup \{i_p\} \setminus \{i_{p+1}\}) - c(S[i_{p+1}]), \quad \forall p = 1, \ldots, r. \tag{26}$$

Therefore, the objective value difference induced by a complete cyclic exchange can be written as

$$\Delta c(S) = \sum_{p=1}^{r} c(S[i_{p+1}] \cup \{i_p\} \setminus \{i_{p+1}\}) - c(S[i_{p+1}]) \tag{27}$$

If $\Delta c(S) < 0$ the according cyclic exchange is called *profitable*. A neighborhood based on cyclic exchanges contains any solution reachable via a cyclic exchange across up to K subsets. Therefore, the number of neighboring solutions is in $\mathcal{O}(n^K)$.

Because of the large neighborhood size the search for an improving neighbor solution cannot be carried out via naive enumeration. In order to allow for a more efficient method the neighborhood is represented by a so-called *improvement graph* $G = (V, E, \delta)$ constructed as follows:

- For each element $a \in A$ a node $v_a \in V$ is created.
- For each valid move of an element $a \in A$ from subset $S[a]$ to subset $S[b]$, $a \neq b$, $S[a] \neq S[b]$, an arc $(v_a, v_b) \in E$ is created.
- With each arc (v_a, v_b) cost $\delta_{v_a, v_b} = c(S[b] \cup \{a\} \setminus \{b\}) - c(S[b])$ are associated.

A cycle $v_{i_1} - v_{i_2} - \ldots - v_{i_r} - v_{i_1}$, $i_p \in A$, $\forall p \in \{1, \ldots, r\}$, is called *subset-disjoint* if $S[i_p] \neq S[i_q]$, $\forall p, q \in \{1 \ldots r\}$, $p \neq q$. A negative-cost subset-disjoint cycle directly corresponds to a profitable cyclic exchange (for a proof see [15]). Although the problem of finding a shortest subset-disjoint cycle in a graph with possibly negative arc costs is $\mathcal{NP}$-hard, a heuristic based on the well-known label-correcting algorithm for finding shortest paths can usually quickly identify good solutions. See Alg. 5.4 for an outline of this procedure as described in [16]. Herein $pred(v)$ denotes the predecessor of node v on the shortest path from any node u to the start node s. $P[u]$ refers to this implicitly defined path and $d(u)$ to the corresponding costs.

The label-correcting algorithm is built upon a data structure $LIST$ which stores nodes having arcs that have yet to be examined. The organization of $LIST$ determines the algorithm's worst-case runtime. Ahuja et al. [16] employ a deque implementation which performs well for sparse graphs [17] even though it leads to exponential worst-case runtime. Because of the dense graphs usually encountered when applying this method to VSLRB, we resort to a FIFO implementation of $LIST$ leading to a worst-case runtime in $\mathcal{O}(|V||E|K)$.

Especially if $|A|$ is large, two major issues have to be considered in practice: (a) the high memory consumption and (b) the computational overhead for the creation of the improvement graph. A possible method to address these problems is to use a different basic set A. In our case we are also able to define

---

Algorithm 5.4. Modified Label-Correcting Algorithm

Input: Improvement graph $G = (V, E, \delta)$, start node $s \in V$
foreach $v \in V \setminus s$ **do**
$\quad\mid\quad d(v) \leftarrow \infty$
$\quad\mid\quad pred(v) \leftarrow null$
$d(s) \leftarrow 0$
$LIST \leftarrow \langle s \rangle$
while $LIST \neq \emptyset$ **do**
$\quad\mid\quad u \leftarrow pop(LIST)$
$\quad\mid\quad$ **if** $P[u]$ *is subset-disjoint* **then**
$\quad\mid\quad\quad\mid\quad$ **foreach** $(u, v) \in E$ **do**
$\quad\mid\quad\quad\mid\quad\quad\mid\quad$ **if** $d(v) > d(u) + \delta_{u,v}$ **then**
$\quad\mid\quad\quad\mid\quad\quad\mid\quad\quad\mid\quad$ **if** $P[u]$ *contains* v **then**
$\quad\mid\quad\quad\mid\quad\quad\mid\quad\quad\mid\quad\quad\mid\quad$ *store subset-disjoint cycle or quit*
$\quad\mid\quad\quad\mid\quad\quad\mid\quad\quad\mid\quad$ **else if** $P[u] \cup v$ *is subset-disjoint* **then**
$\quad\mid\quad\quad\mid\quad\quad\mid\quad\quad\mid\quad\quad\mid\quad d(v) \leftarrow d(u) + \delta_{u,v}$
$\quad\mid\quad\quad\mid\quad\quad\mid\quad\quad\mid\quad\quad\mid\quad pred(v) \leftarrow u$
$\quad\mid\quad\quad\mid\quad\quad\mid\quad\quad\mid\quad\quad\mid\quad LIST \leftarrow LIST \cup v$

---

the improvement graph in terms of replica assignments rather than in terms of requests. As for any assignment $(i, j) \in P$ at most one request is moved in a cyclic exchange both definitions of the improvement graph are equivalent, i.e. they contain the same set of cycles. This definition of the improvement graph is expected to lead to a smaller improvement graph:

Lemma 1. *Let $G_1 = (V_1, E_1, \delta_1)$ and $G_2 = (V_2, E_2, \delta_2)$ denote improvement graphs defined in terms of file requests and replica assignments, respectively. Then the following holds:*

1. $|V_2| \leq |V_1|$
2. $|V_2| < |V_1|$ *if* $\exists i \in F : q_i > m$

Proof. Ad 1.: The first statement obviously holds, because there cannot be more assignments then requests. Ad 2.: Assume that there exists a file with $q_i > m$. Then there must exist at least one assignment $(i, j) \in P$ with $Q(i, j) > 1$. Consequently the number of assignments is smaller than the number requests and therefore $|V_2| < |V_1|$. $\qquad\square$

The basic set A of assignments $(i, j) \in P$, $Q(i, j) > 0$, contains $\mathcal{O}(mn)$ elements. Creating the improvement graph requires enumeration of all pairs of assignments in order to calculate $\mathcal{O}(m^2 n^2)$ arc costs. Calculating the cost of a single arc $(v_{i,j}, v_{f,c})$ and reverting the changes in a naive way requires the four operations **unassign**(f, c), **assign**(i, c), **unassign**(i, c), and **assign**(f, c), leading to a total of $4|E|$ operations. Algorithm 5.5 shows a more efficient way to determine all these arc costs requiring only $2|E| + \mathcal{O}(|A|)$ operations.

Algorithm 5.5. Create VSLRB improvement graph

Input: A solution Q to an instance of VSLRB
foreach $c \in C$ **do**
    foreach $f \in F_c$ **do**
        $V \leftarrow V \cup \{v_{f,c}\}$
        $c_{old} \leftarrow c(S_c)$
        $unassign(f, c)$
        foreach $j \in C \,|\, j \neq c$ **do**
            foreach $i \in F_j \,|\, t_i \in T_c$ **do**
                $V \leftarrow V \cup v_{i,j}$
                $assign(i, c)$
                $c_{new} \leftarrow c(S_c)$
                $E \leftarrow E \cup (v_{i,j}, v_{f,c})$
                $\delta_{v_{i,j}, v_{f,c}} \leftarrow c_{new} - c_{old}$
                $unassign(i, c)$
    $assign(f, c)$

5.5 Neighborhoods of VNS

VNS as depicted in Alg. 5.2 performs shaking by selecting random neighbors from its own neighborhood structures $N_1, \ldots, N_{k_{max}}$ in order to escape from local optima found by the embedded VND. In our case, shaking in a neighborhood N_k is realized by performing k consecutive random moves using the *Access Swap Neighborhood* (see Section 5.2).

6 Experimental Results

In this section, we present representative test results for the proposed MIP and hybrid VNS. We created ten random instances with different characteristics reflecting real-world scenarios. The main characteristics of these test instances are listed in Table 1. Column Z refers to the objective value of the randomly generated initial assignment (i.e. the situation prior to the re-assignment). Three different file types are used: $T = \{\text{Thumbnail}, \text{Preview}, \text{HiRes}\}$. Video runtimes and bitrates b_i were randomly generated using an upper limit of 1800 seconds and 512 kbit/s, respectively. The video file size w_i was derived from these values. The number of expected requests q_i was estimated using a Zipf-like distribution [5] based on randomly assigned video popularities.

Video server characteristics were manually defined. We consider situations with uniform sets $T_j = T, \forall j \in C$, i.e. instances where any server may receive files of any type, as well as situations with non-uniform sets T_j as listed in Table 1. The other server properties U_j, D_j and W_j were chosen uniformly for all instances. U_j and D_j are set to 25 MBit/s except for instances 2 (35 MBit/s), 3 (250 MBit/s), 4 (500 MBit/s), and 5 (250 MBit/s). W_j is set to 180 GB for all instances except for instance 4 where W_j is set to 250 GB. More details can be found in [6]. All test instances are available from the authors upon request.

Table 1. Test instances

Instance	$|C|$	$|F|$	$\sum_{i \in F} q_i$	Z	T_j
1	4	60	489	30710.20	2x {HiRes}, 2x {Preview, Thumbnail}
2	4	300	637	13152.30	uniform
3	5	1200	1328	32844.49	2x {HiRes}, 2x {Preview, HiRes},
					1x {Thumbnail, Preview}
4	7	3000	3064	14492.57	uniform
5	12	4500	4547	24711.20	uniform
6	3	15000	15238	192513.20	1x {HiRes}, 1x {Preview, HiRes}, 1x T
7	20	9000	9027	58700.34	1x {HiRes}, 1x {Preview, HiRes}, 18x T
8	20	3000	3064	31709.60	1x {HiRes}, 1x {Preview, HiRes}, 18x T
9	25	3000	3406	36424.82	1x {HiRes}, 1x {Preview, HiRes}, 23x T
10	25	12000	12680	68269.14	1x {HiRes}, 1x {Preview, HiRes}, 23x T

All tests have been performed on a Linux machine with four 2 GHz dual core AMD Opteron processors and 8 GB RAM. For solving the MIP we used the commercial solver ILOG CPLEX 11.1.

We compare four variants of the VNS: $\text{VNS}_{\text{simple}}$ only includes the simple move and swap neighborhood structures, VNS_{MIP} additionally exploits the κ-server MIP neighborhood structure with $\kappa = 2$, $\text{VNS}_{\text{Cyclic}}$ the cyclic exchange neighborhood structure $\mathcal{N}_{\text{Cyclic}}$, and $\text{VNS}_{\text{MIP+Cyclic}}$ both of them. The weighting factors α and β in the objective function (10) were both set to 1. In the VND, we moved from solution x to solution x' only if the relative objective improvement $\frac{|f(x)-f(x')|}{f(x)}$ was at least 0.01%.

In order to evaluate the performance 30 runs were performed per VNS variant and test instance. Table 2 shows average objective values $\bar{Z}$ of final solutions, corresponding standard deviations σ_Z, and average runtimes $\bar{t}$ for each variant as well as a comparison to results obtained using the MIP approach when using the average $\text{VNS}_{\text{MIP+Cyclic}}$ (or VNS_{MIP}) runtimes as time limits. Column lb further lists the lower bounds obtained from CPLEX.

Even though we employed techniques to reduce the improvement graph size for the cyclic exchange neighborhood (see Sect. 5.4), this data structure became too large to be held in memory for instances 6, 7 and 10. Thus, this neighborhood structure could not be used in these cases.

The pure MIP approach performs well for instances with a limited number of servers. For test instances 1, 2, 3 and 6, which all feature at most four servers the MIP approach produced better results than all of the VNS variants. We exploited this behavior in the κ-Server MIP neighborhood of VNS. The variant $\text{VNS}_{\text{simple}}$ yielded slightly better results than the MIP approach only for instances 5 and 8. The high potential of the two more complex neighborhood structures becomes evident in the case of the large instances 4, 5 and 7 to 10. The best results for these instances were obtained whenever $\mathcal{N}_{\text{Cyclic}}$ was available, either with $\text{VNS}_{\text{MIP+Cyclic}}$ (instances 4 and 5) or $\text{VNS}_{\text{Cyclic}}$ (instances 8 and 9). VNS_{MIP} produces only slightly worse results, but Wilcoxon rank sum tests confirmed

Table 2. Performance comparison of the MIP approach and the four VNS variants

Instance	MIP			VNS											
				VNS_simple			VNS_MIP			VNS_Cyclic			$\text{VNS}_\text{MIP+Cyclic}$		
	Z	lb	$t\,[s]$	$\bar{Z}$	σ_Z	$\bar{t}\,[s]$	$\bar{Z}$	σ_Z	$\bar{t}\,[s]$	$\bar{Z}$	σ_Z	$\bar{t}\,[s]$	$\bar{Z}$	σ_Z	$\bar{t}\,[s]$
1	**1.16**	0.65	3.01	3.17	0.46	0.07	1.82	0.83	1.25	3.05	0.43	0.19	1.60	0.86	2.38
2	**2.07**	0.46	15.01	6.67	0.88	1.28	3.45	0.90	10.43	6.83	0.96	3.72	3.43	1.06	14.67
3	**1.67**	0.17	100.04	3957.78	0.00	0.23	1277.50	1805.29	48.12	2397.71	2398.55	40.40	103.59	388.82	99.87
4	3.07	0.03	141.04	25.49	12.37	5.57	0.48	0.38	22.37	0.49	0.34	330.93	**0.45**	0.34	140.68
5	29.74	0.07	434.11	27.53	9.51	10.54	1.07	0.83	41.65	1.08	0.75	784.59	**0.94**	0.67	434.32
6	**56.10**	54.87	219.04	192522.96	0.00	64.60	67.47	1.53	219.24						
7	639.63	18.62	92.11	12659.38	0.00	37.45	**73.29**	5.41	92.10						
8	133.44	9.80	407.06	50.23	3.36	56.66	43.76	4.93	40.28	**40.89**	4.65	471.86	41.93	3.99	406.47
9	214.09	4.34	731.25	7153.68	523.07	5.85	61.58	6.69	26.49	**55.92**	3.84	1371.99	58.15	5.11	731.12
10	592.96	6.92	175.34	3927.60	632.77	102.75	**83.85**	5.36	174.49						

Table 3. Neighborhood statistics for $\text{VNS}_\text{MIP+Cyclic}$

Instance	$\mathcal{N}_\text{Move}$			$\mathcal{N}_\text{Swap}$			$\mathcal{N}_{2-\text{MIP}}$			$\mathcal{N}_\text{Cyclic}$		
	$\bar{f}$	$\bar{\Delta}$	$\bar{t}\,[s]$	$\bar{f}$	$\bar{\Delta}$	$\bar{t}\,[s]$	$\bar{f}$	$\bar{\Delta}$	$\bar{t}\,[s]$	$\bar{f}$	$\bar{\Delta}$	$\bar{t}\,[s]$
1	9.62	639.79	0.04	46.52	564.07	0.27	5.38	7.82	1.30	0.00	0.00	0.13
2	80.52	2668.80	0.22	135.62	317.81	3.25	10.90	28.23	8.61	0.83	0.05	1.43
3	10.52	741.00	0.14	46.90	765.12	6.88	17.41	3242.91	74.44	12.62	449.58	15.86
4	2.76	206.33	0.19	61.59	508.07	3.59	25.59	1848.41	14.06	1.66	0.72	116.72
5	1.69	83.50	0.27	75.52	371.33	7.60	34.66	2353.52	17.04	2.28	1.42	393.89
6	12.86	3884.09	0.99	73.55	3122.17	5.82	10.90	186175.67	142.12			
7	1.76	15.76	0.60	114.76	1606.29	13.48	59.79	12462.12	35.82			
8	5.14	99.18	0.55	168.52	584.42	79.76	27.79	1974.55	18.04	7.90	2.68	298.83
9	2.21	128.46	0.26	204.41	1892.14	31.18	62.86	8774.38	45.86	23.76	7.66	643.53
10	6.28	576.94	0.94	92.24	1179.68	43.31	60.45	5408.70	53.50			

the significance of these differences with error levels of less than 2.5% on all instances but the first two. An advantage of VNS_{MIP}, however, are its generally considerably shorter runtimes. For instances 7 and 10, when $\mathcal{N}_{Cyclic}$ was not available, the best results also were obtained with VNS_{MIP}.

The bad VNS performance in case of instance 3 is due to its very special structure. In the optimal solution for this instance, only one of several servers accepting the file type *Preview* must be assigned all requests for files of this type. Therefore this particular server must only receive requests without giving off any, which is not achievable using simple swaps or cyclic exchanges. $\mathcal{N}_{2-MIP}$ is not able to relieve this situation either, because of a weakness in the algorithm used to construct the subproblem: Whenever only one server $j \in C$ falls below its target load η_j and at the same time exhibits no overlap in accepted file types with the $\lfloor \frac{\kappa}{2} \rfloor$ heaviest loaded servers, the algorithm cannot determine a valid server selection.

Furthermore, we investigated the contribution of each of the described neighborhood structures in runs of $VNS_{MIP+Cyclic}$. For each instance and each neighborhood structure $\mathcal{N}_{Move}$, $\mathcal{N}_{Swap}$, $\mathcal{N}_{2-MIP}$ and $\mathcal{N}_{Cyclic}$ Table 3 lists the average number of improvements $\bar{f}$, the average total value by which solutions' objective values could be improved $\bar{\Delta}$, and the average time consumed $\bar{t}$. For the majority of the considered test instances $\mathcal{N}_{2-MIP}$ turned out to be the most effective neighborhood structure in terms of total improvement as well as in terms of consumed runtime. Nonetheless, $\mathcal{N}_{Cyclic}$ was still capable of achieving further improvements at the cost of significantly larger runtimes.

7   Conclusion and Future Work

In this work we presented two approaches for solving a particular *Video-Server Load Re-Balancing* (VSLRB) problem. First, we described a MIP formulation which we solved by a general purpose MIP solver. This approach is able to identify high-quality solutions for problem instances involving a small number of servers. For solving larger instances in a better way, we developed a VNS with an embedded VND. Besides the simple move and swap neighborhood structures, two more sophisticated large neighborhood search methods are included: The benefits of the MIP-approach are exploited in the κ-Server MIP neighborhood, and a variant of a cyclic exchange neighborhood, adapted to cope with very large improvement graphs, is searched by an efficient label-correcting shortest path algorithm. On average, the VNS approach was able to identify substantially better solutions than the MIP approach for all of the six test instances involving more than five servers. Both large neighborhood methods are able to dramatically boost the performance of the simple VNS variant, although the additional contributions of $\mathcal{N}_{Cyclic}$ are (naturally) rather small when applied in conjunction with the MIP-based neighborhood search. Future work might address certain weaknesses with special scenarios like the one illustrated with test instance 3 by considering further neighborhood structures.

References

1. Ghose, D., Kim, H.: Scheduling Video Streams in Video-on-Demand Systems: A Survey. Multimedia Tools and Applications 11(2), 167–195 (2000)
2. Dan, A., Sitaram, D., Shahabuddin, P.: Scheduling policies for an on-demand video server with batching. In: Proceedings of the second ACM international conference on Multimedia, pp. 15–23. ACM, New York (1994)
3. Venkatasubramanian, N., Ramanathan, S.: Load management in distributed video servers. In: Proceedings of the 17th International Conference on Distributed Computing Systems (ICDCS 1997), Washington, DC, USA, p. 528. IEEE Computer Society, Los Alamitos (1997)
4. Wolf, J., Yu, P., Shachnai, H.: Disk load balancing for video-on-demand systems. Multimedia Systems 5(6), 358–370 (1997)
5. Zhou, X., Xu, C.: Optimal Video Replication and Placement on a Cluster of Video-on-Demand Servers. In: Proceedings of the International Conference on Parallel Processing, Washington, DC, USA, pp. 547–555. IEEE Computer Society Press, Los Alamitos (2002)
6. Walla, J.: Exakte und heuristische Optimierungsmethoden zur Lösung von Video Server Load Re-Balancing. Master's thesis, Vienna University of Technology, Vienna, Austria (2009)
7. Yu, H., Zheng, D., Zhao, B.Y., Zheng, W.: Understanding user behavior in large-scale video-on-demand systems. In: Proceedings of the 1st ACM SIGOPS/EuroSys European Conference on Computer Systems 2006 (EuroSys 2006), pp. 333–344. ACM, New York (2006)
8. Cherkasova, L., Gupta, M.: Analysis of enterprise media server workloads: access patterns, locality, content evolution, and rates of change. IEEE/ACM Transactions on Networking 12(5), 781–794 (2004)
9. Griwodz, C., Bär, M., Wolf, L.: Long-term movie popularity models in video-on-demand systems: or the life of an on-demand movie. In: Proceedings of the fifth ACM international conference on Multimedia, pp. 349–357. ACM, New York (1997)
10. Chen, K., Chen, H.-C., Borie, R., Liu, J.C.L.: File replication in video on demand services. In: Proceedings of the 43rd annual ACM Southeast Regional Conference (ACM-SE 43), pp. 162–167. ACM, New York (2005)
11. Wang, Y., Liu, J., Du, D., Hsieh, J.: Efficient video file allocation schemes for video-on-demand services. Multimedia Systems 5(5), 283–296 (1997)
12. Aggarwal, G., Motwani, R., Zhu, A.: The load rebalancing problem. In: Proceedings of the fifteenth annual ACM symposium on Parallel algorithms and architectures, pp. 258–265. ACM, New York (2003)
13. Allahverdi, A., Ng, C., Cheng, T., Kovalyov, M.: A survey of scheduling problems with setup times or costs. European Journal of Operational Research 187(3), 985–1032 (2008)
14. Hansen, P., Mladenović, N.: Variable Neighbourhood Search. In: Glover, Kochenberger (eds.) Handbook of Metaheuristics, pp. 145–184. Kluwer Academic Publisher, New York (2003)
15. Thompson, P., Orlin, J.: The theory of cyclic transfers. Operations Research Center Working Papers. Massachusetts Institute of Technology (1989)
16. Ahuja, R., Orlin, J., Sharma, D.: New Neighborhood Search Structures for the Capacitated Minimum Spanning Tree Problem. Sloan School of Management, Massachusetts Institute of Technology (1998)
17. Bertsekas, D.P.: A simple and fast label correcting algorithm for shortest paths. Networks 23(7), 703–709 (1993)

Effective Hybrid Stochastic Local Search Algorithms for Biobjective Permutation Flowshop Scheduling

Jérémie Dubois-Lacoste, Manuel López-Ibáñez, and Thomas Stützle

IRIDIA, CoDE, Université Libre de Bruxelles, Brussels, Belgium
jeremie.dl@gmail.com, {manuel.lopez-ibanez,stuetzle}@ulb.ac.be

Abstract. This paper presents the steps followed in the design of hybrid stochastic local search algorithms for biobjective permutation flow shop scheduling problems. In particular, this paper tackles the three pairwise combinations of the objectives (i) makespan, (ii) the sum of the completion times of the jobs, and (iii) the weighted total tardiness of all jobs. The proposed algorithms are combinations of two local search methods: two-phase local search and Pareto local search. The design of the algorithms is based on a careful experimental analysis of crucial algorithmic components of the two search methods. The final results show that the newly developed algorithms reach very high performance: The solutions obtained frequently improve upon the best nondominated solutions previously known, while requiring much shorter computation times.

1   Introduction

In this paper, we tackle biobjective flowshop scheduling problems. The flowshop environment models problems where each job consists of a set of operations that are to be carried on machines, and the machine order is the same for each job. Flowshops are a common production environment, for example in the chemical or the ceramic tile industry. We consider flowshops minimizing the following criteria: the completion time of the last job (makespan), which has been the most intensively studied criterion for this problem [1]; the sum of completion times (total flowtime) of all jobs, which recently has attracted a lot of efforts [2,3]; and the weighted tardiness, a criterion which is important in practical applications [4]. For an overview of the biobjective flowshop problems that result from each combination of these objectives, we refer to Minella et al. [5].

At a high level, our approach is based on the hypothesis that effective stochastic local search (SLS) algorithms for multi-objective combinatorial optimization problems (MCOPs) can be obtained by (i) developing (or simply using known) very effective algorithms for the underlying single-objective problems, and (ii) using these single-objective algorithms as components of higher-level algorithm frameworks for tackling multi-objective problems. Further, multi-objective specific local search routines may be examined as an alternative for reaching high-performance algorithms or as a post-processing step.

M.J. Blesa et al. (Eds.): HM 2009, LNCS 5818, pp. 100–114, 2009.

As a first step in our SLS algorithm development, we adopted the state-of-the-art SLS algorithm for the flowshop problem under makespan objective, the iterated greedy (IG) algorithm of Ruiz and Stützle [6]. Subsequently, this algorithm was extended to the sum of flowtime and weighted tardiness objectives and we fine-tuned the resulting algorithms. In a next step, we extended these algorithms to tackle the biobjective versions of the flowshop problem that result from each of the pairwise combinations of the three above mentioned objectives. This was done by integrating the IG algorithms into the two-phase local search (TPLS) framework [7]. At the same time, we also implemented Pareto local search (PLS) [8] algorithms that use different neighborhood structures. A core part of our work is the careful experimental study of the main algorithmic components of the resulting TPLS and PLS algorithms. The insights from this study were then used to propose a hybrid SLS algorithm that combines the TPLS and the PLS algorithms. The final experimental results with this algorithm show its excellent performance: It often finds better Pareto fronts than those of a reference set that was extracted from the best nondominated solutions obtained by a set of 23 other algorithms.

The paper is structured as follows. In Section 2 we introduce basic notions needed in the following. In Section 3 we describe the single-objective algorithms that underlie the two-phase local search approach. Section 4 presents results of the various experimental studies and we conclude in Section 5.

2   Preliminaries

2.1   Multi-objective Optimization

In MCOPs, (candidate) solutions are ranked according to an *objective function vector* $\boldsymbol{f} = (f_1, \ldots, f_d)$ with d objectives. If no *a priori* assumptions upon the decision maker's preferences can be made, the goal typically becomes to determine a set of feasible solutions that "minimize" $\boldsymbol{f}$ in the sense of Pareto optimality. If $\boldsymbol{u}$ and $\boldsymbol{v}$ are vectors in $\mathbb{R}^d$, we say that $\boldsymbol{u}$ *dominates* $\boldsymbol{v}$ ($\boldsymbol{u} \prec \boldsymbol{v}$) iff $\boldsymbol{u} \neq \boldsymbol{v}$ and $u_i \leq v_i$, $i = 1, \ldots, d$; we say that $\boldsymbol{u}$ *weakly dominates* $\boldsymbol{v}$ ($\boldsymbol{u} \leq \boldsymbol{v}$) iff $u_i \leq v_i$, $i = 1, \ldots, d$. We also say that $\boldsymbol{u}$ and $\boldsymbol{v}$ are *nondominated* iff $\boldsymbol{u} \nprec \boldsymbol{v}$ and $\boldsymbol{v} \nprec \boldsymbol{u}$ and are (pairwise) *non weakly dominated* if $\boldsymbol{u} \nleq \boldsymbol{v}$ and $\boldsymbol{v} \nleq \boldsymbol{u}$. For simplicity, we also say that a solution s dominates another one s' iff $\boldsymbol{f}(s) \prec \boldsymbol{f}(s')$. If no other s' exists such that $\boldsymbol{f}(s') \prec \boldsymbol{f}(s)$, the solution s is called a *Pareto optimum*. The goal in MCOPs typically is to determine the set of all Pareto optimal solutions. Since this task is in many cases computationally intractable, in practice the goal becomes to find an approximation to the set of Pareto optimal solutions in a given amount of time that is as good as possible. In fact, any set of mutually nondominated solutions provides such an approximation. The notion of Pareto optimality can be extended to compare sets of mutually nondominated solutions [9]. In particular, we can say that one set A dominates another set B ($A \prec B$), iff every $\boldsymbol{b} \in B$ is dominated by at least one $\boldsymbol{a} \in A$.

2.2   Bi-objective Permutation Flowshop Scheduling

In the flowshop scheduling problem (FSP) a set of n jobs $(J_1, \ldots, J_n)$ is given to be processed on m machines $(M_1, \ldots, M_m)$. All jobs go through the machines in the same order, i.e., all jobs have to be processed first on machine M_1, then on machine M_2 and so on until machine M_m. A common restriction in the FSP is to forbid job passing, i.e., the processing sequence of the jobs is the same on all machines. In this case, candidate solutions correspond to permutations of the jobs and the resulting problem, on which we focus here, is the permutation flowshop scheduling problem (PFSP). All processing times p_{ij} for a job J_i on a machine M_j are fixed, known in advance and nonnegative. In the following, we denote by C_i the completion time of a job J_i on machine M_m. For a given job permutation π, the makespan is the completion time of the last job in the permutation, i.e., $C_{\max} = C_{\pi(n)}$. For $m \geq 3$ this problem is $\mathcal{NP}$-hard in the strong sense [10]. In the following, we refer to this problem as $PFSP\text{-}C_{\max}$.

The other objectives we study are the minimization of the sum of flowtimes and the minimization of the weighted tardiness. The sum of flowtimes is defined as $\sum_{i=1}^{n} C_i$. The resulting PFSP with this objective is strongly $\mathcal{NP}$-hard even with only two machines [10]. We refer to this problem as $PFSP\text{-}SFT$. For the weighted tardiness objective, each job has a due date d_i by which it is to be finished and a weight w_i indicating its priority. The tardiness is defined as $T_i = \max\{C_i - d_i, 0\}$ and the total weighted tardiness is given by $\sum_{i=1}^{n} w_i \cdot T_i$. This problem we denote $PFSP\text{-}WT$; it is strongly $\mathcal{NP}$-hard even for a single machine.

In this paper, we tackle the three biobjective problems that result from the three possible pairs of objectives. A number of algorithms have been proposed to tackle each of these biobjective problems, but rarely more than one possible combination of the objectives has been addressed in a paper. The algorithmic approaches range from constructive algorithms to applications of SLS methods such as evolutionary algorithms, tabu search, or simulated annealing. Minella et al. [5] give a comprehensive overview of the literature on the three problems we tackle here and present the results of an extensive experimental analysis of 23 algorithms, either specific or adapted for tackling the three biobjective PFSPs. They identify MOSA [11] as the best performing algorithm.

2.3   Two-Phase Local Search and Pareto Local Search

In this paper, we study SLS algorithms that represent two main classes of multi-objective SLS algorithms [12]: algorithms that follow a component-wise acceptance criterion (CWAC), and those that follow a scalarized acceptance criterion (SAC). As two paradigmatic examples of each of these classes, we use two-phase local search (TPLS) [7] and Pareto local search (PLS) [8].

Two-Phase Local Search. The first phase of TPLS uses an effective single-objective algorithm to find a good solution for one objective. This solution is the initial solution for the second phase, where a sequence of scalarizations are solved by an SLS algorithm. Each scalarization transforms the multi-objective problem into a single-objective one using a weighted sum aggregation. For a given weight

---

Algorithm 1. Two-Phase Local Search

Input: A random or heuristic solution π
$\pi' := SLS_1(\pi)$;
for all weight vectors $\boldsymbol{\lambda}$ **do**
    $\pi' := SLS_2(\pi', \boldsymbol{\lambda})$;
    Add π' to Archive;
end for
Filter Archive;

---

vector $\boldsymbol{\lambda} = (\lambda_1, \lambda_2)$, the value w of a solution s with objective function vector $\boldsymbol{f}(s) = (y_1, y_2)$ is computed as $w = (\lambda_1 \cdot y_1) + (\lambda_2 \cdot y_2)$, s.t. $\lambda_1, \lambda_2 \in [0, 1] \subset \mathbb{R}$ and $\lambda_1 + \lambda_2 = 1$. In TPLS, each run of the SLS algorithm for solving a scalarization uses as an initial solution the best one found for the previous scalarization. The motivation for using such a method is to exploit the effectiveness of the underlying single-objective algorithm. Algorithm 1 gives the pseudocode of TPLS. We denote by SLS_1 the SLS algorithm to minimize the first single objective. SLS_2 is the SLS algorithm to minimize the weighted sums.

Pareto Local Search. PLS is an iterative improvement method for solving MCOPs that is obtained by replacing the usual acceptance criterion of iterative improvement algorithms for single-objective problems by an acceptance criterion that uses the dominance relation. Given an initial archive of unvisited nondominated solutions, PLS iteratively applies the following steps. First, it randomly chooses an unvisited solution s from the candidate set. Then, the neighborhood of s is fully explored and all neighbors that are not weakly dominated by s or by any solution in the archive are added to the archive. Solutions in the archive dominated by the newly added solutions are eliminated. Once the neighborhood of s is fully explored, s is marked as visited. The algorithm stops when all solutions in the archive have been visited.

We also implemented the *component-wise step* (CW-step) procedure as a postprocessing step of the solutions produced by TPLS. It adds nondominated solutions in the neighborhood of the solutions returned by TPLS to the archive, but it does not explore the neighborhood of these newly added solutions further. Hence, CW-step may be interpreted as a specific variant of PLS with an early stopping criterion. Because of this early stopping criterion, the CW-step results in worse nondominated sets than PLS. However, compared to running a full PLS, CW-step typically requires a very small additional computation time.

3 Single-Objective SLS Algorithms

The performance of the single-objective algorithms used by TPLS is crucial. They should be state-of-the-art algorithms for the underlying single-objective problems and as good as possible for the scalarized problems resulting from the weighted sum aggregations. Motivated by these considerations, for $PFSP\text{-}C_{\max}$ we reimplemented in C++ the iterated greedy (IG) algorithm ($IG\text{-}C_{\max}$) by Ruiz

---

Algorithm 2. Iterated Greedy

$\pi := \text{NEH}$;
while termination criterion not satisfied **do**
    $\pi_R := Destruction(\pi)$;
    $\pi' := Construction(\pi_R)$;
    $\pi' := LocalSearch(\pi')$ % optional;
    $\pi := AcceptanceCriterion(\pi, \pi')$;
end while

---

and Stützle [6], which is a current state-of-the-art algorithm for this problem. An algorithmic outline is given in Algorithm 2. The essential idea of IG is to iterate over a construction heuristic by first destructing partially a complete solution; next, from the resulting partial solution π_R a full problem solution is reconstructed and possibly further improved by a local search algorithm. This solution is then accepted in dependence of an acceptance criterion.

More concretely, IG-$C_{\max}$ uses the NEH heuristic [13] for constructing the initial solution and for reconstructing full solutions in the main IG loop. (NEH is an insertion heuristic that sorts the jobs according to some criterion and inserts jobs in this order into the partial schedule. Note that this sorting is only relevant when NEH constructs the initial solution; in the main loop of IG the jobs are considered in random order.) In the destruction phase a small number of d randomly chosen jobs are removed. The local search is an effective first-improvement algorithm based on the *insert* neighborhood, where two solutions are neighbors if they can be obtained by removing a job from one position and inserting it in a different one. The acceptance criterion uses the Metropolis condition: A worse solution is accepted with a probability given by $\exp\left(f(\pi') - f(\pi)\right)/T$, where f is the objective function and the temperature parameter T is maintained constant throughout the run of the algorithm. Parameter values are given in Table 1.

Given the known very good performance of IG-$C_{\max}$, we use it also for the other two objectives. However, the speed-ups of Taillard for $C_{\max}$ [14] are not anymore applicable, which leads to a factor n increase of the local search time complexity. As a side result, it is unclear whether the same neighborhood as for the makespan criterion should be chosen. We have therefore considered also (i) the exchange neighborhood, where two solutions are neighbors if they can be obtained by exchanging the position of two jobs; and (ii) the swap neighborhood, where only two adjacent jobs are exchanged. We tested only restricted versions of the insert and exchange neighborhoods, where the possible insertion and exchange moves of only one job are examined.

Other changes concern the formula for the definition of the temperature parameter for the acceptance criterion. This is rather straightforward for *PFSP-SFT*, which can be done by adapting slightly the way the temperature is defined. For *PFSP-WT* no input data-driven setting as for the other two objectives could be obtained due to large variation of the objective values. Therefore, the temperature parameter is defined relating it to a given target percentage deviation from the current solution. Finally, for *PFSP-WT* we explored different ways of defining the initial sequence of jobs for the NEH heuristic.

Table 1. Adaptation of IG for each objective and for the scalarized problems from each combination of objectives. A ($\downarrow$) denotes a decreasing order, and a ($\uparrow$) denotes an increasing order. Tp is a parameter of the formula for Temperature. Settings for *IG-$C_{\max}$* follow [6]. For *IG-SFT*, the formula of Temperature is the same as for *IG-$C_{\max}$* but multiplied by n. For *IG-WT*, the initial order for NEH is given by the well-known SLACK heuristic [4] augmented with priorities w_i. Parameter d is the number of jobs removed in the destruction phase. Insert-T refers to a full insert iterative improvement using the speed-ups of Taillard [14]; Swap to a full iterative improvement execution using the swap neighborhood and Ins. to the insertion search for one job. For details see the text.

Algorithm	Init. order for NEH	Temperature T	Tp	d	LS
IG-$C_{\max}$	$\sum_{j=1}^{m} p_{ij}$ ($\downarrow$)	$Tp \cdot \frac{\sum_{i=1}^{n}\sum_{j=1}^{m} p_{ij}}{n \cdot m \cdot 10}$	0.4	4	Insert-T
IG-SFT	$\sum_{j=1}^{m} p_{ij}$ ($\downarrow$)	$Tp \cdot \frac{\sum_{i=1}^{n}\sum_{j=1}^{m} p_{ij}}{m \cdot 10}$	0.5	5	Swap
IG-WT	$w_i \cdot (d_i - C_i(s))$ ($\uparrow$)	$\frac{100}{Tp \cdot f(s)}$	0.7	4	Swap+Ins.
IG-($C_{\max}$, SFT)	$\sum_{j=1}^{m} p_{ij}$ ($\downarrow$)	$\frac{100}{Tp \cdot f(s)}$	0.5	5	Swap
IG-($\cdot$, WT)	$w_i \cdot (d_i - C_i(s))$ ($\uparrow$)	$\frac{100}{Tp \cdot f(s)}$	0.5	5	Swap

We tuned the IG algorithms for *PFSP-SFT* and *PFSP-WT* using iterated F-Race [15] on training instances that are different from those used for the final test results. The final configurations retained from this tuning phase are given in Table 1. The lines *IG-($C_{\max}$, SFT)* and *IG-($\cdot$, WT)* concern the scalarized problems where the weights are different from one and zero for the indicated objectives. A closer examination of the performance of the resulting single-objective algorithms (not reported here) showed that for total flowtime the final IG algorithm is competitive to current state-of-the-art algorithms as of 2009; for the total tardiness objective the performance is also very good and very close to state-of-the-art; in fact we could improve with the IG algorithms a large fraction (in each case more than 50%) of the best known solutions of available benchmark sets.

4   Multi-Objective SLS Algorithms

In what follows, we first study main algorithm components of the PLS and TPLS algorithms and then present a comparison of a final hybrid SLS algorithm to reference sets of the best solutions found so far for a number of benchmark instances. We used the benchmark from Minella et al. [5], which consists of the benchmark set of Taillard [16] augmented with due dates and priorities. In order to avoid over-tuning, we performed the algorithm component analysis on 20 training instances of size 50x20 and 100x20, which were generated following the procedure used by Minella et al. [5].

The results are analyzed by graphically examining the attainment surfaces of a single algorithm and differences between the empirical attainment functions

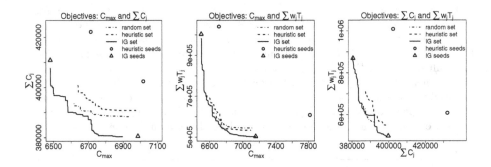

Fig. 1. Nondominated sets obtained by PLS using different quality of seeds for instance 100x20_3. The randomly generated solution is outside the range shown.

(EAF) of pairs of algorithms. The EAF of an algorithm provides the probability, estimated from several runs, of an arbitrary point in the objective space being attained (weakly dominated) by a solution obtained by a single run of the algorithm [17]. An attainment surface delimits the region of the objective space attained by an algorithm with a certain minimum probability. In particular, the worst attainment surface delimits the region of the objective space always attained by an algorithm, whereas the best attainment surface delimits the region attained with the minimum non-zero probability. Similarly, the median attainment surface delimits the region of the objective space attained by half of the runs of the algorithm. Examining the attainment surfaces allows to assess the likely location of the output of an algorithm. On the other hand, examining the differences between the EAFs of two algorithms allows to identify regions of the objective space where one algorithm performs better than another. Given a pair of algorithms, the differences in favor of each algorithm are plotted side-by-side and the magnitude of the difference is encoded in gray levels. López-Ibáñez et al. [18] provide a detailed explanation of these graphical techniques.

4.1 Analysis of PLS Components

Seeding. As a first experiment, we analyzed the computation time required and the final quality of the nondominated sets obtained by PLS when PLS is seeded with solutions of different quality. We test seeding PLS with: (i) one randomly generated solution, (ii) two good solutions (one for each single objective) obtained by the NEH heuristics (see Table 1), and (iii) two solutions obtained by IG for each objective after 10 000 iterations. Figure 1 gives representative examples of nondominated sets obtained by PLS for each test and indicates the initial seeding solutions of NEH and IG. The best nondominated sets, in terms of a wider range of the Pareto front and higher quality solutions, are obtained when using the IG seeds. Generally, seeding PLS with very good solutions produces better nondominated sets; this result is strongest for the biobjective problem that considers makespan and total flowtime. We further examined the

Table 2. Computation time of PLS for different types of seeds

Objectives	Instance Size	random avg.	sd.	heuristic avg.	sd.	IG avg.	sd.
$(C_{max}, \sum C_i)$	50x20	8.85	2.05	6.23	2.48	4.56	0.38
	100x20	177.40	27.60	142.23	29.79	162.14	26.09
$(C_{max}, \sum w_i T_i)$	50x20	31.61	6.84	33.85	7.46	24.02	3.84
	100x20	641.96	215.55	767.23	299.33	626.48	114.08
$(\sum C_i, \sum w_i T_i)$	50x20	26.72	3.02	28.17	2.62	23.70	3.33
	100x20	742.42	157.10	807.75	121.70	895.23	176.29

Table 3. Computation time of PLS for different neighborhood operators

Objectives	Instance Size	exchange avg.	sd.	insertion avg.	sd.	ex. + ins. avg.	sd.
$(C_{max}, \sum C_i)$	50x20	2.21	0.35	1.57	0.44	4.84	1.06
	100x20	77.56	19.44	70.91	12.8	157.64	30.26
$(C_{max}, \sum w_i T_i)$	50x20	12.94	3.11	10.11	1.75	23.03	4.09
	100x20	314.63	69.08	251.84	49.33	611.6	115.02
$(\sum C_i, \sum w_i T_i)$	50x20	14.24	3.79	9.51	1.8	23.72	3.87
	100x20	492.91	102.59	239.04	101.47	872.32	262.21

computation time required by PLS in dependence of the initial seed in Table 2. Our conclusion is that seeding PLS with very good initial solutions does not strongly reduce computation time. However, given the strong improvement on solution quality, seeding PLS with solutions obtained by TPLS is pertinent.

Neighborhood operator. We experiment with PLS variants using three neighborhoods: (i) insertion, (ii) exchange, and (iii) exchange plus insertion. The latter simply checks for all moves in the exchange and insertion neighborhood of each solution. We measured the computation time of PLS with each underlying operator for each combination of objectives (Table 3). The computation time of the combined exchange and insertion neighborhood is slightly more than the sum of the computation times for the exchange and insertion neighborhoods. For comparing the quality of the results, we examine the EAF differences of 10 independent runs. Figure 2 gives two representative examples. Typically, the exchange and insertion neighborhoods lead to better performance in different regions of the Pareto front (top plot), and both of them are consistently outperformed by the combined exchange and insertion neighborhood (bottom plot).

4.2 Analysis of TPLS Components

Number of scalarizations and number of iterations. In TPLS, each scalarization is computed using a different weight vector. In this paper, we use a regular

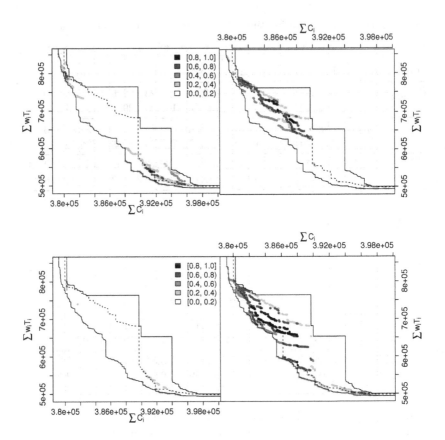

Fig. 2. EAF differences for (*top*) insertion vs. exchange and (*bottom*) exchange vs. exchange and insertion. The combination of objectives is $\sum C_i$ and $\sum w_i T_i$. Dashed lines are the median attainment surfaces of each algorithm. Black lines correspond to the overall best and overall worst attainment surfaces of both algorithms.

sequence of weight vectors from $\lambda = (1, 0)$ to $\lambda = (0, 1)$. If N_{scalar} is the number of scalarizations, the successive scalarizations are defined by weight vectors $\lambda_i = (1 - (i/N_{\text{scalar}}), i/N_{\text{scalar}}), i = 0, \ldots, N_{\text{scalar}}$.

For a fixed computation time, in TPLS there is a tradeoff between the number of scalarizations to be used and the number of iterations to be given for each of the invocations of the single-objective SLS algorithm. In fact, the number of scalarizations (N_{scalar}) determines how many scalarized problems will be solved (intuitively, the more the better approximations to the Pareto front may be obtained), while the number of iterations (N_{iter}) of IG determines decisively how good the final IG solution will be. Here, we examine the trade-off between the settings of these two parameters by testing all 9 combinations of the following settings: $N_{\text{scalar}} = \{10, 31, 100\}$ and $N_{\text{iter}} = \{100, 1\,000, 10\,000\}$.

We first studied the impact of increasing either N_{scalar} or N_{iter} for a fixed setting of the other parameter. Although clear improvements are obtained by

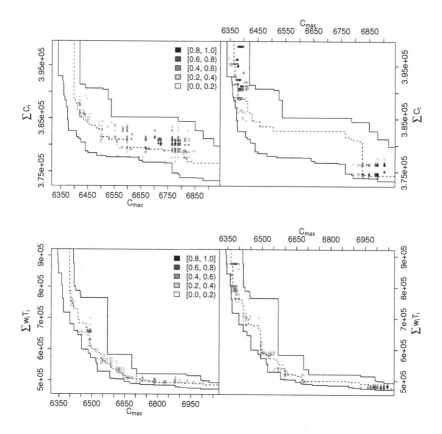

Fig. 3. EAF differences between $N_{scalar} = 100$ and $N_{iter} = 1000$, versus $N_{scalar} = 10$ and $N_{iter} = 10000$ for two combinations of objectives: C_{max} and $\sum C_i$ (top) and C_{max} and $\sum w_i T_i$ (bottom)

increasing each of the two parameters, there are significant differences. While for the number of scalarizations some type of limiting behavior without strong improvements was observed when going from 31 to 100 scalarizations (while improvements from 10 to 31 were considerable), increasing the number of iterations of IG alone seems always to produce significant improvements.

Next, we compare settings that require roughly the same computation time. Figure 3 compares a configuration of TPLS using $N_{scalar} = 100$ and $N_{iter} = 1000$ against other configuration using $N_{scalar} = 10$ and $N_{iter} = 10000$. Results are shown for two of the three combinations of objective functions. (The results are representative for other instances.) As illustrated by the plots, there is no clear winner in this case. A larger number of iterations typically produces better solutions in the extremes of the Pareto front. On the other hand, a larger number of scalarizations allows to find trade-off solutions that are not found with a smaller number of scalarizations. Given these results, among settings that require

Table 4. Average computation time and standard deviation for CW-step and PLS

Objectives	Instance Size	CW-step avg.	sd.	PLS avg.	sd.
$(C_{\max}, \sum C_i)$	50x20	0.20	0.02	2.40	0.75
	100x20	1.56	0.40	75.04	32.53
$(C_{\max}, \sum w_i T_i)$	50x20	0.37	0.03	7.43	2.11
	100x20	2.51	0.42	202.71	50.51
$(\sum C_i, \sum w_i T_i)$	50x20	0.34	0.04	8.99	1.71
	100x20	2.75	0.34	373.15	87.44

roughly the same computation time, there is no single combination of settings that produces the best results overall among all objectives and instances.

Double TPLS. We denote as Double TPLS (DTPLS) the following strategy. First, the scalarizations go sequentially from one objective to the other one, as in the usual TPLS. Then, another sequence of scalarizations is performed starting from the second objective back to the first one. To introduce more variability, the weight vectors used in the first TPLS pass are alternated with the weight vectors used for the second TPLS pass. We compared this DTPLS strategy with the simple TPLS using 30 scalarizations of 1000 iterations. Although we found on several instances strong differences, these differences were not consistent in favor of advantages of DTPLS over TPLS or vice versa. This gives some evidence that the two strategies do not behave the same, but we left it for further research to investigate which instances features may explain the observed differences.

TPLS + PLS vs. TPLS + CW-step. As a final step, we compare the performance tradeoffs incurred by either running PLS or the CW-step to the archive of solutions returned by TPLS. For all instances, we generated 10 initial sets of solutions by running TPLS with 30 scalarizations of 1000 iterations each. Then, we independently apply to these initial sets the CW-step and PLS, both using the exchange and the insertion neighborhoods. In other words, each method starts from the same set of initial solutions in order to reduce variance.

Table 4 gives the additional computation time that is incurred by PLS and the CW-step after TPLS has finished. The results clearly show that the CW-step incurs only a minor overhead with respect to TPLS, while PLS requires considerably longer times, especially on larger instances. Moreover, the times required to terminate PLS are much lower than when seeding it with only two very good solutions (compare with Table 2). With respect to solution quality, Figure 4 compares TPLS versus TPLS+CW-step (top), and TPLS+CW step versus TPLS+PLS (bottom). As expected, the CW-step is able to slightly improve the results of TPLS, while PLS produces much better results. In summary, if computation time is very limited, the CW-step provides significantly better results at almost no computational cost; however, if enough time is available, PLS improves much further than the sole application of the CW-step.

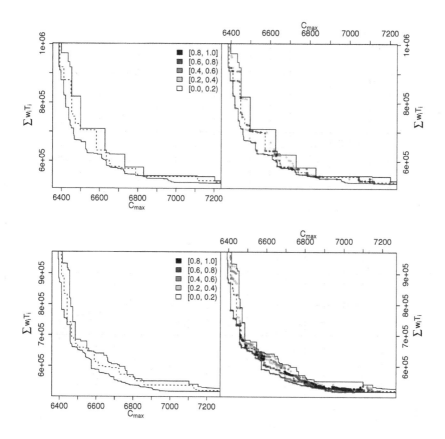

Fig. 4. EAF differences between (top) simple TPLS vs. TPLS + CW-step, and (bottom) TPLS + CW-step vs. TPLS + PLS. Objectives are C_{max} and $\sum w_i T_i$.

4.3   Comparison to Existing Algorithms

In order to compare our algorithm with existing work, we used the benchmark of Minella et al. [5], which is Taillard's benchmark set augmented with due dates. In their review, the authors compare 23 heuristics and metaheuristics using biobjective combinations of makespan (C_{max}), sum of flowtime ($\sum C_i$), and total tardiness ($\sum T_i$). They also provide the best-known nondominated solutions obtained across 10 runs of each of the 23 algorithms. We use these sets as reference sets to compare with our algorithm. As the reference sets are given for the total (not weighted) tardiness criterion, we slightly modified our algorithm by setting all the priorities to one ($w_i = 1$).

In particular, we compare our results with the reference sets given for a computation time of 200 seconds and corresponding to instances from `ta081` to `ta090` (size 100x20). These reference sets were obtained on an Intel Dual Core E6600 CPU running at 2.4 Ghz. By comparison, our algorithms were run on a Intel Xeon E5410 CPU running at 2.33 Ghz with 6MB of cache size, under Cluster

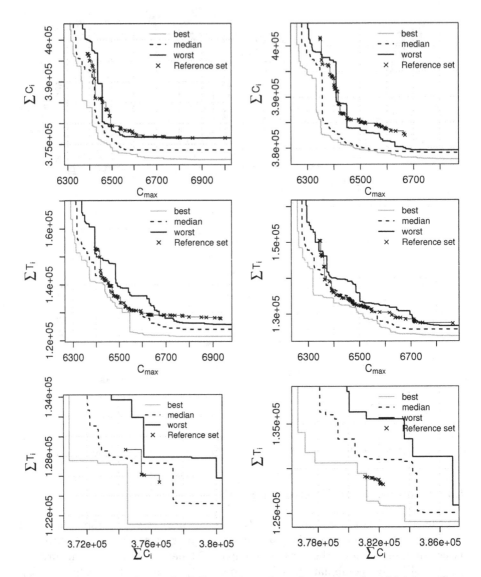

Fig. 5. Comparison of our algorithm against the reference set for objectives $C_{\max}$ and $\sum C_i$ (top), $C_{\max}$ and $\sum T_i$ (middle), and $\sum C_i$ and $\sum T_i$ (bottom) on instances DD_Ta081 (left) and DD_Ta082 (right)

Rocks Linux. Both machines result in approximately similar speed, however, to be conservative, we decided to round down the quality of our results by using only 150 CPU seconds. For our algorithms, we used the following parameter settings. The two extreme solutions are generated by running IG for 10 seconds each. Then TPLS starts from the solution obtained for the first objective and runs 14 scalarizations of 5 seconds each. Finally, we apply PLS in the exchange and insertion neighborhoods and stop it after 60 CPU seconds. We repeat each run 10 times with different random seeds.

For each instance, we compare the best, median and worst attainment surfaces obtained by our algorithm with the corresponding reference set. Figure 5 shows results for the three objective combinations. In most cases, the median attainment surface of our algorithm is very close to (and often dominates) the reference set obtained by 10 runs of 23 algorithms, each run using 200 CPU seconds. Moreover, the current state-of-the-art algorithms for these problems are among these algorithms. Therefore, we conclude that our algorithm is clearly competitive and probably superior to the current state-of-the-art for these problems. All the additional plots are available online at http://iridia.ulb.ac.be/supp/IridiaSupp2009-004

5   Conclusions

In this paper, we have studied algorithmic components of the TPLS and PLS algorithms for three biobjective permutation flowshop problems, and we proposed a hybrid, high-performing SLS algorithm for these problems.

The final experimental results have shown that our SLS algorithms are able to significantly improve upon the reference sets of the nondominated solutions that have been obtained during an extensive experimental study of 23 algorithms for the same biobjective problems. These and other recent results in the literature [7,19] suggest that hybrid algorithms combining the TPLS and PLS frameworks have a large potential to improve upon the current state-of-the-art in multi-objective optimization.

Acknowledgments. This work was supported by the META-X project, an *Action de Recherche Concertée* funded by the Scientific Research Directorate of the French Community of Belgium. Thomas Stützle acknowledges support from the Belgian F.R.S.-FNRS, of which he is a Research Associate.

References

1. Ruiz, R., Maroto, C.: A comprehensive review and evaluation of permutation flow-shop heuristics. European Journal of Operational Research 165, 479–494 (2005)
2. Liao, C.J., Tseng, C.T., Luarn, P.: A discrete version of particle swarm optimization for flowshop scheduling problems. Computers & Operations Research 34(10), 3099–3111 (2007)
3. Tsenga, L.Y., Lin, Y.T.: A hybrid genetic local search algorithm for the permutation flowshop scheduling problem. European Journal of Operational Research 198(1), 84–92 (2009)
4. Vallada, E., Ruiz, R., Minella, G.: Minimising total tardiness in the m-machine flowshop problem: A review and evaluation of heuristics and metaheuristics. Computers & Operations Research 35(4), 1350–1373 (2008)
5. Minella, G., Ruiz, R., Ciavotta, M.: A review and evaluation of multiobjective algorithms for the flowshop scheduling problem. INFORMS Journal on Computing 20(3), 451–471 (2008)

6. Ruiz, R., Stützle, T.: A simple and effective iterated greedy algorithm for the permutation flowshop scheduling problem. European Journal of Operational Research 177(3), 2033–2049 (2007)
7. Paquete, L., Stützle, T.: A two-phase local search for the biobjective traveling salesman problem. In: Fonseca, C.M., Fleming, P.J., Zitzler, E., Deb, K., Thiele, L. (eds.) EMO 2003. LNCS, vol. 2632, pp. 479–493. Springer, Heidelberg (2003)
8. Paquete, L., Chiarandini, M., Stützle, T.: Pareto local optimum sets in the biobjective traveling salesman problem: An experimental study. In: Gandibleux, X., et al. (eds.) Metaheuristics for Multiobjective Optimisation. LNEMS, vol. 535, pp. 177–200. Springer, Heidelberg (2004)
9. Zitzler, E., Thiele, L., Laumanns, M., Fonseca, C.M., Grunert da Fonseca, V.: Performance assessment of multiobjective optimizers: An analysis and review. IEEE Transactions on Evolutionary Computation 7(2), 117–132 (2003)
10. Garey, M.R., Johnson, D.S., Sethi, R.: The complexity of flowshop and jobshop scheduling. Mathematics of Operations Research 1, 117–129 (1976)
11. Varadharajan, T.K., Rajendran, C.: A multi-objective simulated-annealing algorithm for scheduling in flowshops to minimize the makespan and total flowtime of jobs. European Journal of Operational Research 167(3), 772–795 (2005)
12. Paquete, L., Stützle, T.: Stochastic local search algorithms for multiobjective combinatorial optimization: A review. In: Gonzalez, T.F. (ed.) Handbook of Approximation Algorithms and Metaheuristics. Chapman & Hall/CRC (2007); 29–1—29–15
13. Nawaz, M., Enscore Jr., E., Ham, I.: A heuristic algorithm for the m-machine, n-job flow-shop sequencing problem. OMEGA 11(1), 91–95 (1983)
14. Taillard, É.D.: Some efficient heuristic methods for the flow shop sequencing problem. European Journal of Operational Research 47(1), 65–74 (1990)
15. Balaprakash, P., Birattari, M., Stützle, T.: Improvement strategies for the F-Race algorithm: Sampling design and iterative refinement. In: Bartz-Beielstein, T., Blesa Aguilera, M.J., Blum, C., Naujoks, B., Roli, A., Rudolph, G., Sampels, M. (eds.) HM 2007. LNCS, vol. 4771, pp. 108–122. Springer, Heidelberg (2007)
16. Taillard, É.D.: Benchmarks for basic scheduling problems. European Journal of Operational Research 64(2), 278–285 (1993)
17. Grunert da Fonseca, V., Fonseca, C.M., Hall, A.: Inferential performance assessment of stochastic optimisers and the attainment function. In: Zitzler, E., Deb, K., Thiele, L., Coello Coello, C.A., Corne, D.W. (eds.) EMO 2001. LNCS, vol. 1993, pp. 213–225. Springer, Heidelberg (2001)
18. López-Ibáñez, M., Paquete, L., Stützle, T.: Exploratory analysis of stochastic local search algorithms in biobjective optimization. Technical Report TR/IRIDIA/ 2009-015, IRIDIA, Université Libre de Bruxelles, Brussels, Belgium (May 2009)
19. Lust, T., Teghem, J.: Two-phase Pareto local search for the biobjective traveling salesman problem. Journal of Heuristics (to appear, 2009), doi:10.1007/s10732-009-9103-9

Hierarchical Iterated Local Search for the Quadratic Assignment Problem

Mohamed Saifullah Hussin and Thomas Stützle

IRIDIA, Université Libre de Bruxelles, Brussels, Belgium
{mbinhuss,stuetzle}@ulb.ac.be

Abstract. Iterated local search is a stochastic local search (SLS) method that combines a perturbation step with an embedded local search algorithm. In this article, we propose a new way of hybridizing iterated local search. It consists in using an iterated local search as the embedded local search algorithm inside another iterated local search. This nesting of local searches and iterated local searches can be further iterated, leading to a hierarchy of iterated local searches. In this paper, we experimentally examine this idea applying it to the quadratic assignment problem. Experimental results on large, structured instances show that the hierarchical iterated local search can offer advantages over using a "flat" iterated local search and make it a promising technique to be further considered for other applications.

1 Introduction

Hybrid stochastic local search (SLS) algorithm are those that combine various other SLS strategies into one more complex SLS algorithm [1]. In this sense, iterated local search (ILS) is such a hybrid SLS method since it combines local search (one strategy) with perturbations (another strategy) that are typically of a different nature than the modifications applied in the local search [2]. Following another nomenclature, ILS is also classified as being one (typically non-hybrid) metaheuristic. One goal of the research area of *hybrid metaheuristics* is to study high-level strategies that combine *different* metaheuristics into a new metaheuristic [3]. If, hence, the goal is to propose a hybrid metaheuristic, in the case of ILS this is a rather trivial undertaking. If we use inside ILS instead of an iterative improvement local search a tabu search or a simulated annealing, two examples of other metaheuristics, we already have a hybrid metaheuristic. In fact, such proposals have studied [4,5,6,7]. Recently, also hybrids between ILS and evolutionary algorithms have been considered, where the evolutionary algorithm plays the role of the perturbation operator in ILS [8].

In this article, we propose a new way of generating a hybrid "metaheuristic": we hybridize a metaheuristic with itself! To be more concrete, here we propose to hybridize ILS with ILS. In fact, this is can be accomplished by using inside one ILS algorithm a local search procedure that is another ILS algorithm. Such a type of hybridization can actually make sense. In fact, many hybrid algorithms

M.J. Blesa et al. (Eds.): HM 2009, LNCS 5818, pp. 115–129, 2009.

are obtained by using some higher level "perturbation mechanism" and combining this with a local search type algorithm. This is the case, for example, for hybrid algorithms that combine evolutionary algorithms with local search algorithms (resulting in so called memetic algorithms) or that combine ant colony optimization with local search algorithms. Note that in these cases, the local search part is usually taken by iterative improvement algorithms but also many examples exist of using tabu search, simulated annealing or even iterated local search [9]. In our proposed ILS-ILS hybrid, the outer ILS algorithm plays this role of the perturbation mechanism while the inner ILS algorithm plays the role of the local search.

Interestingly, this idea of nesting iterated local searches can be further iterated, by hybridizing ILS with an ILS-ILS hybrid and so on. Hence, we can define a hierarchy of nested ILS algorithms. Given this possibility, we have denoted this class of hybrids as hierarchical ILS (HILS). In fact, the ILS-ILS hybrid is simply the first level of this hierarchy with one ILS on top and one in the next lower level; therefore, we denote this hybrid as HILS(2); HILS(1) is a flat ILS without any nested ILS process.

In this paper, we explore this idea of hierarchical ILS applying it to the quadratic assignment problem (QAP). In particular, we build an HILS(2) algorithm using two particular ILS algorithms as the underlying building blocks. The computational results obtained with the resulting HILS(2) algorithm are very promising. On a number of instances, a tuned version of HILS(2) performs significantly better than tuned versions of the underlying ILS algorithms. Some limited experiments with an untuned HILS(3) algorithm show that the consideration of further levels of this hierarchy may be interesting for QAP instances with a particular structure.

This article is structured as follows. In the next section, we give an overview of ILS, introduce the hierarchical ILS approach in more detail, and illustrate in Section 3 its adaptation to the QAP. In Section 4 we introduce the benchmark instances used in this article. Experimental results with the hierarchical ILS are given in Section 5 and we conclude in Section 6.

2   Hierarchical Iterated Local Search

Iterated Local Search (ILS) iterates over a heuristic, in the following called *local search*, by building a chain of local optima. It builds this chain by perturbing at each iteration a current local optimum; this results in an intermediate solution, which is the starting solution of the next local search. Once the local search terminates, the new local optimum is accepted depending on some acceptance test. An algorithmic outline of ILS is given in Algorithm 1. ILS starts building this chain of solutions using some initial solution and applying a local search to it. The goal of the perturbation is to modify a current local optimum to generate a good new starting solution for a local search. The perturbation should not be too strong, since otherwise the perturbation is close to a random restart; but also not too weak, since otherwise the following local search could simply undo

Algorithm 1. Iterated Local Search

---

procedure *ILS(s₀)*
input: s_0 % initial solution
$s^* = Local\_Search(s_0)$
while termination condition is not met **do**
    $s' = Perturbation(s^*, history)$
    $s^{*'} = Local\_Search(s')$
    $s^* = Acceptance\_Criterion(s^*, s^{*'}, history)$
end while
return s_b^* % best solution found in the search process
end *Iterated Local Search*

---

the changes and return to the previous local optimum. The acceptance criterion decisively determines how greedy the search is. For a more detailed discussion of ILS we refer to [2].

The role of the *Local_Search* procedure can be played by any heuristic that takes as input some solutions $s \in S$, where S is the space of all candidate solutions for the problem under concern, and outputs an improved solution s^*. In fact, for the following discussion only this input–output behavior is important, that is, we can see local search as a function *Local_Search* : $S \mapsto S^*$; we call $S^* \subseteq S$, for simplicity, the space of "local optima". If the local search is an iterative improvement algorithm, S^* is the space of local optima w.r.t. the neighborhood structure used in the local search; if the local search is a tabu search, then S^* is the set of the best solutions found by a tabu search execution of a specific search length and using a specific s as input. This point of view results in the interpretation of ILS as being a stochastic search in the space of local optima, the space of "local optima" being defined by the output behavior of the embedded heuristic.

The underlying idea of the hierarchical ILS is to use two (or more) nested iterated local searches. The first level of this hierarchy is obtained by replacing the procedure *Local_Search* in Algorithm 1 by an ILS algorithm. This results in an HILS(2) that is outlined in Algorithm 2. This outline corresponds to the first hierarchy level of HILS and it will be denoted by HILS(2), since it consists of two nested ILS algorithms.

When applying the above interpretation of ILS as searching the space of local optima of *Local_Search* to HILS(2), we get the following relationship. The procedure *ILS* can be interpreted as a function *ILS* : $S^* \mapsto S^{**}$, that is, it is simply defining a mapping from the space of local optima S^* to a smaller space S^{**}. We have that $S^{**} \subseteq S^*$ and it can be interpreted as the set of possible outputs of the inner ILS, that is, it is the set of local optima that correspond to the potential best solutions that are found by the inner ILS. The "outer" ILS is, hence, doing a stochastic search in a potentially even smaller space than S^*. Clearly, this hierarchy could be further extended by considering further levels, leading to an interpretation of HILS as searching in S^*, S^{**}, S^{***}, and so on, depending on the number of nested iterated local searches. In this article, we examine in some more detail an HILS(2) algorithm; in addition, we provide some evidence

---

Algorithm 2. Hierarchical Iterated Local Search; level 2

---

procedure *Iterated Local Search*
$s_0 = Generate\_Initial\_Solution$
$s^{**} = ILS(s_0)$
while termination condition is not met **do**
   $s' = Perturbation(s^{**}, history)$
   $s^{**\prime} = ILS(s')$
   $s^{**} = Acceptance\_Criterion(s^{**}, s^{**\prime}, history)$
end while
return s_b^{**}, the best solution found by HILS
end *Iterated Local Search*

---

that at least for some class of QAPLIB instances an HILS(3) algorithm shows very good performance.

3 Hierarchical ILS for the Quadratic Assignment Problem

3.1 Quadratic Assignment Problem

The QAP is an NP-hard problem [10] that has attracted a large number of research efforts [11,12]. It is an abstract model of many practical problems arising in applications such as hospital, factory or keyboard layout. In the QAP, one is given two $n \times n$ matrices A and B, where a_{ij} gives the distance between the pair of positions i and j; b_{kl} gives the flow of, for example, patients or material between units k and l. Assigning units k and l to positions i and j, respectively, incurs a cost contribution of $a_{ij} \cdot b_{kl}$. The goal of the QAP is to assign units to positions such that the sum of all cost contributions is minimized. A candidate solution for the QAP can be represented as a permutation π, where $\pi(i)$ is the unit that is assigned to position i. The objective function of the QAP is

$$f(\pi) = \sum_{i=1}^{n} \sum_{j=1}^{n} a_{ij} \cdot b_{\pi(i)\pi(j)}. \tag{1}$$

Due to its difficulty for exact algorithms, currently SLS algorithms define the state-of-the-art for the QAP [1]. ILS algorithms have shown very good performance on the QAP, some variants reaching state-of-the-art performance [13].

3.2 Iterated Local Search for the Quadratic Assignment Problem

Various ILS algorithms for the QAP have been studied in [13] and we describe here the main ILS components taken from [13] for our study.

Our ILS and HILS algorithms start from a randomly generated initial solution. As the local search we consider an iterative improvement algorithm in the two-exchange neighborhood using a first improvement pivoting rule. The two-exchange neighborhood $\mathcal{N}(\pi)$ of a solution π is given by the set of permutations that can be obtained by exchanging units r and s at positions π_r and π_s. As soon as an improving move is found during the exchange, it is immediately applied. To avoid spending too much time during the search, the iterative improvement algorithm exploits the *don't look bits* technique to speed up the local search algorithm. If no improvement is found during the neighborhood scan of an unit, the corresponding *don't look bit* is turned on (that is, set to one) and the unit is not considered as a starting unit for the next iteration. If an unit is involved in a move and changes its position, its *don't look bit* is turned off again (that is, set to zero). The purpose of using *don't look bits* is to restrict the attention to the most interesting part of the local search.

The perturbation in ILS and HILS exchanges k randomly chosen units. We consider various possibilities for defining the strength of the perturbation, that is, the number k of units exchanged and consider two specific schemes: keeping the number k fixed or using a variable perturbation strength, as done in Variable Neighborhood Search (VNS) [14].

The main differences among the ILS algorithms studied by Stützle [13] concerned the acceptance criteria. In particular, he studied the better acceptance criterion (*Better*), where a new local optimum $s^{*\prime}$ is accepted if it is better than the incumbent one s^*, a random walk acceptance criterion (*RandomWalk*), where a new local optimum $s^{*\prime}$ is accepted irrespective of its quality (that is, always), and a restart acceptance criterion (*Restart*), which behaves like *Better* with one exception: if for i_r successive iterations no improved solution has been found, the ILS algorithm is restarted from a new, randomly generated solution.

Note that in [13] an ILS algorithm based on a Simulated Annealing type acceptance criterion (*LSMC*) was also studied. We did not use this one here, mainly to simplify the construction of an appropriate HILS algorithm by needing to adapt less parameters. Another reason is that for most QAP instances, the best results of the ILS algorithms that use the *Better*, *RandomWalk*, and *Restart* acceptance criteria are roughly on par or better than those of *LSMC* [13]. Similarly, the two population-based ILS variants replace worst (*RepWorst*) and evolution strategies (*ES*) are not considered as building blocks here, since the target here is an ILS variant that does not use a population.

3.3   Hierarchical Iterated Local Search

For defining an HILS(2) algorithm, we need to define the acceptance criteria and perturbations used at each of the hierarchy levels—the local search algorithm is only used in the inner ILS.

The construction of the HILS(2) algorithm is based on the following main rationale. The inner ILS should provide a fast search for good solutions in a confined area of the search space, while the outer ILS loop is responsible for guiding the search towards appropriate regions of the search space. We translate this into the outer ILS algorithm doing rather large jumps in the search space, corresponding to large perturbations, while the inner ILS algorithm is only using small perturbations.

The only remaining component to be set is the acceptance criterion. In this article, we consider for both hierarchy levels only two possibilities: either using *Better* or *RandomWalk*. This results in a total of four possible combinations. We compare the HILS(2) algorithm also to ILS with *Restart* acceptance criterion; note that this comparison is particularly interesting, since ILS with restarts can be seen as an extreme case of HILS: it corresponds to an outer ILS loop that uses a perturbation size equal to the instance size (that is, it generates a new random solution) and the *RandomWalk* acceptance criterion.

Finally, note that in HILS(2) one additional parameter arises: the execution time given to the inner ILS algorithm. In fact, this type of parameter arises in any similar hybrid where a metaheuristic is used as an improvement method and we will set this parameter empirically.

4   Benchmark Instances

While most of the studies of algorithms for the QAP are based on instances from QAPLIB, in this study we use benchmark instances that we have randomly generated. There are a few reasons for this. A first is that in this way we can systematically vary instance characteristics. Secondly, in this way we can generate a larger set of reasonably uniform instances, which can be used, for example, for automated tuning tools and allow for the comparison on test instances. Thirdly, this allows also to study the algorithm behavior on larger instances, since most QAPLIB instances are of rather small size.

In this study, we generated a new set of QAP instances similar to some types of instances used in the research of the Metaheuristics Network [15]. These instances are generated such that they have widely different distributions of the entries of the flow matrices. In particular, the flow matrices are generated such that their entries are similar to the type of distributions found in real-world QAP instances [16].

The distance matrices are generated using two possibilities: Euclidean distances and grid distances.

Euclidean Distances. Using this approach, the distance matrix entries are generated as the pairwise Euclidean distances between n points in the plane:
  1. each coordinate of each of n points is generated randomly according to a uniform distribution in $[0, k]$.
  2. The Euclidean distance between each pair of two points is calculated, rounded to the nearest integer, and returned as the distance matrix entry.
Grid Distances. In this approach, the distance matrix corresponds to the Manhattan distance between the points on a $A \times B$ grid.

Instances generated based on Euclidean distances will have most probably only one single optimal solution, while instances generated based on Manhattan distances have at least four optimal solutions, due to the symmetries in the distance matrix.

Table 1. Basic statistics on some instance classes of the benchmark instances. Given are the distance dominance (dd), the flow dominance (fd), and the sparsity of the flow matrix (sp).

Class	size	dd	fd	sp	Class	size	dd	fd	sp
ES.1	100	47.17	274.57	0.20	GS.1	100	69.31	275.37	0.21
ES.2	100	47.17	386.28	0.45	GS.2	100	69.31	385.57	0.43
ES.3	100	47.15	493.25	0.90	GS.3	100	69.31	492.93	0.89
ES.1	200	47.41	276.27	0.21	GS.1	200	68.98	276.42	0.20
ES.2	200	47.41	386.59	0.45	GS.2	200	68.98	386.26	0.42
ES.3	200	47.42	496.29	0.91	GS.3	200	68.98	494.12	0.92

Flow matrix entries follow an exponential distribution, where a large number of entries have small values, and very few are with large values. The flow matrices of all instances are symmetric and have a null diagonal.

Structured Flows. Instances with structured flows are generated as follows:

1. n points are generated randomly according to a uniform distribution in a square with dimension 100×100.
2. If the distance between two points i and j is above a threshold value p, the flow is zero.
3. Otherwise, the flow is calculated as $x = -1 \cdot \log(R)$, where R is a random number in $[0, \cdots, 100]$, and $flow = \min\left\{100 \cdot x^{2.5}, 10000\right\}$

The numbers have been chosen in such a way that the objective values can still be represented as unsigned integers on a 32-Bit system.

As a result, we have two different levels of distance matrices (Euclidean (ES) and grid distances (GS)). For our experiments, we have generated for each level of distance matrices three different levels of flow matrices. Each of these three levels of flow matrices differs mainly in their sparsity (defined as the percentage of zero entries) and the resulting flow dominance values (which corresponds to the variation coefficient of the flow matrix entries multiplied by 100). Statistics on these values (including also the distance dominance) for each resulting instance class are given in Table 1. Of each of these classes, we have generated 100 instances for tuning and 100 for testing. The statistics are given for two instance sizes.

5   Experimental Results

5.1   Experimental Setup

All the experiments that we report here have been run on Intel Xeon 2.4 Ghz quad-core CPUs with 6 MB cache and 8 GB of RAM running under Cluster Rocks Linux. Due to the sequential implementation of the algorithms only a single core is used for computations. If nothing else is said, the stopping criteria for the experiments on the various instance classes were based on the CPU time that is taken by our reimplementation of the robust tabu search algorithm of Taillard [17] to perform $1000 \cdot n$ iterations. The algorithms are coded in C and compiled with gcc version 3.4.6.

5.2   Choice of Local Search

As a first step, we identified the most appropriate local search to be used in ILS and HILS(2). To do so, we used a set of 30 instances of size 100 and we performed tests considering the following four options:

1. first-improvement two-exchange local search without don't look bits.
2. first-improvement two-exchange local search with don't look bits.
3. first-improvement two-exchange local search with don't look bits and resetting of don't look bits during perturbation.
4. best-improvement two-exchange local search.

Since ILS is the main part of HILS(2), the performance of local search was studied using exclusively an ILS algorithm. A statistical analysis of the experimental results showed that the third option was the most performing one. In this option, after a perturbation only the don't look bits of the units that have changed the position are reset to zero.

5.3   Experimental Study of HILS Combinations

As a next step, we generated systematically a number of HILS(2) variants. In total, eight variants have been studied that were obtained by all possible combinations of the following settings. The outer and the inner ILS can use the acceptance criteria *Better* or *RandomWalk* and the length of the inner ILS is terminated after either n or $2n$ ILS iterations, where n is the instance size. The perturbation sizes were set, following the rationale given in Section 3.3, as follows. The outer ILS used variable perturbation sizes taken from the set $\{0.25n, 0.5n, 0.75n\}$, while the inner ILS varied the perturbation size among the integers in the interval $[3, 0.25n - 1]$; in both cases, the changes in the perturbation sizes follow the rules of the basic VNS [14].

Each of the resulting eight HILS(2) algorithms was applied to 100 instances of the above mentioned benchmark set of sizes 100 and 300. The results of the experiments were analyzed using pairwise Wilcoxon tests using Holm's correction to identify the winning combinations. While we expected to observe clear trends concerning the winning configurations, no fully conclusive results were obtained. In fact, seven of the eight configurations have been among the best performing configurations (taken into account statistically not significantly different performance). The only slight trend appeared to be that on less sparse instances, for the outer ILS the *Better* acceptance criterion appeared to be preferable, while for very sparse instances the *randomWalk* criterion gave overall better performance.

5.4   Performance Comparison of HILS and ILS Variants

Given that no fully conclusive results were obtained in our previous analysis, as a next step, we fine-tuned two ILS algorithms that were used as building blocks of HILS(2) (ILS with better acceptance criterion (ILSb), ILS with random walk acceptance criterion (ILSrw)), as well as the ILS algorithm with *Restart* acceptance criterion (ILSrr). For this task, we used F-race, a racing algorithm based on the Friedman two-way analysis of variance [18]; for each tuning task a same maximum number of 6 000 algorithm evaluations was given as the computational budget for F-race.

The tuning of ILSb, ILSrw, and ILSrr was considered to allow for a fair comparison. For the first two variants, only one parameter was tuned; the perturbation scheme, while for ILSrr, an additional parameter, the number of iterations without improvement before a random restart is triggered (i_r), also needs to be tuned. Perturbation schemes considered are a fixed perturbation scheme and variable neighborhood search (VNS). For each scheme, several choices of perturbation strength (k), step size ($k+$), and maximum perturbation strength (k_{max}) are available. The winning configurations from F-Race are then used when comparing HILS(2) to the underlying ILS variants. Table 2 shows the parameter settings selected by F-Race for the three basic ILS variants.

Four parameters were considered for tuning HILS(2); the perturbation scheme, the acceptance criterion for the outer loop (Acc_o) and the inner loops (Acc_i), and the length of the inner loop (L) as a multiple of the instance size. The perturbation scheme consists of several subsidiary parameters; perturbation direction P, which controls how the perturbation size changes (increasing, decreasing, or fix), minimum perturbation strength for the outer loop, k_{min-o}, change in perturbation strength after every iteration $k+$, maximum perturbation strength for outer loop, k_{max-o}, and maximum perturbation strength for inner loop, k_{max-i}. Acceptance criterion Acc_o and Acc_i consist each of two choices as explained previously: *Better* (b) and *Random Walk* (rw). The parameter settings of HILS(2) chosen by F-Race is summarized in Table 3.

On most of ES instances, the fixed perturbation scheme was performing best, while for GS instances, the VNS scheme seems to be the better choice. Intuitively, if a fixed perturbation scheme is chosen, the perturbation strength has to be very large to allow stronger diversification by the algorithm. Results from F-Race confirm this since the surviving perturbation strengths for a fixed perturbation size are always large: $0.7n$ or $0.9n$. For the VNS scheme, several k_{min-o} values were considered during tuning, and all the values, $n/6$, $n/5$, and $n/4$ are chosen as good parameter settings for respective instance class. F-Race suggests a long inner loop length for ES instances, while for GS instances, a shorter inner loop length was chosen all the time.

The results of the comparison between HILS(2) and the three ILS algorithms are presented next. For each combination of instance class and size, we first computed the average solution quality across the 100 test instances and we used them as the basis of our comparison. The percentage difference between these

Table 2. Tuned parameter settings for basic ILS variants. The two entries for each parameter correspond to instances of size 100 and 200. See the text for more details.

Instances	Pert. 100	Pert. 200	k_{min} 100	k_{min} 200	$k+$ 100	$k+$ 200	k_{max} 100	k_{max} 200	ir 100	ir 200
					ILSb					
ES.1	fix	fix	40	80	0	0	40	80	-	-
ES.2	fix	fix	40	80	0	0	40	80	-	-
ES.3	VNS	VNS	3	3	3	3	90	180	-	-
GS.1	VNS	VNS	3	3	3	1	90	180	-	-
GS.2	VNS	fix	3	20	1	0	90	20	-	-
GS.3	VNS	VNS	3	3	3	1	90	180	-	-
					ILSrw					
ES.1	VNS	VNS	3	3	3	3	90	180	-	-
ES.2	VNS	VNS	3	3	3	3	90	180	-	-
ES.3	VNS	fix	3	140	1	0	30	140	-	-
GS.1	VNS	fix	3	20	3	0	90	20	-	-
GS.2	VNS	fix	3	20	3	0	90	20	-	-
GS.3	fix	fix	40	140	0	0	40	140	-	-
					ILSrr					
ES.1	fix	fix	10	80	0	0	10	80	1	1
ES.2	fix	fix	10	80	0	0	10	80	1	2
ES.3	VNS	VNS	3	3	3	3	90	180	1	1
GS.1	VNS	fix	3	80	3	0	90	80	1	1
GS.2	VNS	VNS	3	3	3	1	90	90	1	1
GS.3	VNS	VNS	3	3	1	1	90	90	10	2

Table 3. Parameter settings for HILS(2). The two entries for each parameter correspond to instances of size 100 and 200. See the text for more details and an explanation of the table entries.

Parameter Instances	P 100	P 200	k_{min-o} 100	k_{min-o} 200	$k+$ 100	$k+$ 200	k_{max-o} 100	k_{max-o} 200	k_{max2} 100	k_{max2} 200	Acc_o 100	Acc_o 200	Acc_i 100	Acc_i 200	L 100	L 200
ES.1	VNS	fix	25	140	25	0	75	140	25	140	rw	b	rw	rw	30	30
ES.2	fix	fix	70	140	0	0	70	140	70	140	rw	b	rw	rw	30	10
ES.3	fix	fix	90	180	0	0	90	180	90	180	b	b	rw	rw	10	2
GS.1	VNS	VNS	17	34	17	34	83	167	17	34	b	rw	b	rw	2	2
GS.2	VNS	VNS	20	50	20	50	80	150	20	50	rw	rw	b	rw	2	5
GS.3	VNS	fix	17	180	17	0	83	180	17	180	b	b	b	rw	2	5

values and results obtained by other algorithms are then computed. In Table 4 we give the results obtained by HILS(2) and the underlying ILS variants. Pairwise Wilcoxon tests with Holm corrections for multiple comparisons were conducted to check the statistical significance of our results. If the p-value is lower than our chosen level of $\alpha = 0.05$, the difference between the corresponding algorithms is deemed significant, and printed in italics.

Table 4. Comparison between HILS, ILSb, ILSrw, and ILSrr. Given are the percentage deviations from the best average across the 100 test instances. Statistically significant differences are indicated in italics font. See the text for details.

Class	Size	t	HILS	ILSb	ILSrw	ILSrr
ES.1	100	54	0.00	*0.33*	0.00	*0.22*
ES.2	100	54	0.00	*0.43*	*0.06*	*0.21*
ES.3	100	54	0.00	*1.07*	*0.94*	0.07
GS.1	100	54	0.00	*0.08*	*0.02*	*0.01*
GS.2	100	54	0.00	*0.11*	*0.02*	*0.02*
GS.3	100	54	0.00	*0.03*	*0.26*	*0.02*
ES.1	200	438	0.00	*0.26*	0.02	*0.23*
ES.2	200	438	0.00	*0.27*	0.00	*0.24*
ES.3	200	438	0.00	*0.38*	*0.49*	0.07
GS.1	200	438	0.00	*0.26*	*0.08*	*0.14*
GS.2	200	438	0.00	*0.45*	*0.05*	*0.02*
GS.3	200	438	0.00	0.02	*0.33*	0.68

The result of this experiment show very promising performance of HILS(2). HILS(2) has obtained the lowest average for 11 out of 12 instance classes; in many cases the observed differences are statistically significant. Only on dense ES instance set of size 100, the average obtained by ILS$_{rw}$ is very slightly less than that of HILS(2) (the truncation to two significant positions hides this fact in the table), but not in a statistically significant way.

5.5 Comparison of HILS, ILS-ES, ILSts, and RoTS

As a next step in our analysis, we compared the performance of HILS(2) to three well known algorithms from literature. The first is robust tabu search (RoTS) by Taillard [17], an algorithm, which is chosen as a baseline in many comparisons of SLS algorithms, iterated tabu search (ILSts), variants of which have been discussed in several articles [4,5,6,7], and a state-of-the-art population-based ILS variant, the ILS-ES from [13]. All algorithms are re-implemented and run for a same computation time. The results of the comparison are given in Table 5. As in the previous comparison, we normalize the results w.r.t. the best performing algorithm and indicate significantly worse performance according to the Wilcoxon test with Holm's corrections for multiple comparisons in italics font. On instance size 100, ILS-ES is clearly the best performing algorithm: it obtains significantly better solutions on every instance class considered. For larger instances, HILS(2) shows a particularly good performance on dense instances. Only on very sparse instances (those with class identifier 3), ILS-ES shows (slightly) better average results. This behavior is true on instance sizes 200, 300, and 500.

5.6 HILS(3)

As a final experiment, we conducted tests on structured Taillard's instances (taiXXe instances) of size 125, 175, and 343 [19]. These instances were designed

Table 5. Comparison between HILS(2), ILS-ES, ILSts, and RoTS. The computation times for each instance class and all algorithms are indicated in the column *t*. Given are the percentage deviations from the best average across the 100 test instances. Statistically significant differences are indicated in italics font.

Class	*Size*	*t*	HILS(2)	ILS-ES	ILSts	RoTS
ES.1	100	54	*0.08*	0.00	*0.24*	*0.26*
ES.2	100	54	*0.06*	0.00	*0.26*	*0.26*
ES.3	100	54	*0.27*	0.00	*0.30*	*0.95*
GS.1	100	54	*0.03*	0.00	*1.99*	*0.29*
GS.2	100	54	*0.02*	0.00	*1.79*	*0.28*
GS.3	100	54	*0.01*	0.00	*5.16*	*0.42*
ES.1	200	438	0.00	*0.11*	*0.11*	*0.20*
ES.2	200	438	0.00	*0.09*	*0.14*	*0.17*
ES.3	200	438	*0.27*	0.00	*1.31*	*0.94*
GS.1	200	438	0.00	0.01	*0.65*	*0.34*
GS.2	200	438	0.00	0.03	*0.66*	*0.30*
GS.3	200	438	*0.07*	0.00	*1.17*	*0.72*
ES.1	300	1400	0.00	*0.17*	*0.08*	*0.22*
ES.2	300	1400	0.00	*0.16*	*0.17*	*0.24*
ES.3	300	1400	*0.32*	0.00	*1.15*	*0.77*
GS.1	300	1400	0.00	*0.13*	*1.04*	*0.39*
GS.2	300	1400	0.00	*0.14*	*0.99*	*0.28*
GS.3	300	1400	*0.06*	0.00	*2.52*	*1.53*
ES.1	500	7300	0.00	*0.48*	*0.54*	*0.64*
ES.2	500	7300	0.00	*0.53*	*0.36*	*0.67*
ES.3	500	7300	*0.19*	0.00	*0.55*	*3.25*
GS.1	500	7300	0.00	*1.70*	*6.95*	*5.27*
GS.2	500	7300	0.00	*2.02*	*6.87*	*5.20*
GS.3	500	7300	*0.02*	0.00	*2.91*	*2.18*

Table 6. Comparison between HILS(2), HILS(3), ILS-ES, and RoTS on three taiXXe instances. Given are the percentage deviations from the best average for each test instance.

Instance	HILS(2)	HILS(3)	ILS-ES	RoTS
tai125e01	157.00	0.00	1.00	25.00
tai175e01	220.00	0.00	2.00	23.00
tai343e01	59.00	0.00	4.00	34.00

to be hard for local search algorithms by integrating some block structures in the flow and distance matrices. We selected one instance per size and conducted 20 independent trials for each algorithm. Since we found rather poor behavior of HILS(2) on this class of instances, on which it is known that large perturbations are needed, we tested also an ad-hoc variant of an HILS(3) algorithm. This HILS(3) was using at each of the three hierarchy levels a *Better* acceptance criterion. The perturbation size was at each level modified in a VNS fashion using as

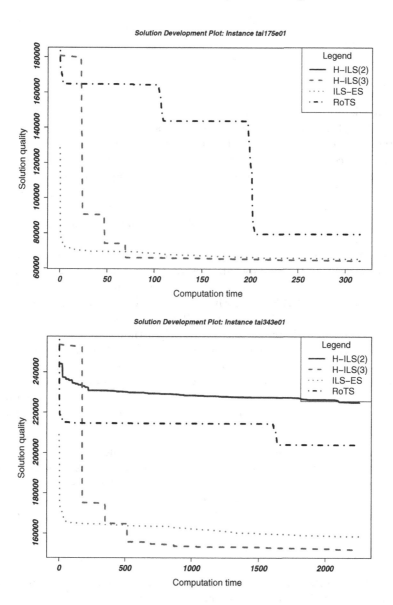

Fig. 1. Plot of the development of the solution cost over computation time on two taiXXe instances. Given are the plots for HILS(2), HILS(3), ILS-ES and robust tabu search (RoTS).

possible perturbation strengths in the outermost ILS the set $\{0.25n, 0.5n, 0.75n\}$, in the middle-level ILS the set $\{11, 12, .., 0.25n\}$, and in the innermost level the set $\{3, 4, .., 10\}$. The innermost ILS was terminated after ten iterations, while the middle-level ILS was terminated after five iterations. As can be seen in Table 6, HILS(3) resulted to be the best performing algorithm for these instances, even performing better than the ILS-ES approach. An illustration of the development of the solution quality over time is given in Figure 1.

6   Conclusions

In this article, we have introduced an example of a metaheuristic that has been successfully hybridized with itself, which shows that hybrid metaheuristics are potentially not only limited to hybrids among *different* metaheuristics. We illustrated this concept of self-hybridization using the example of iterated local search, where this concept can be applied in a reasonably straightforward way. In fact, the possibility of having nested local searches was already suggested in an earlier overview article on ILS [2], although there the idea of a self-hybridization of ILS has not been further elaborated. We further have shown that this idea of nesting iterated local searches leads to a hierarchy of ILS algorithms.

In an experimental study, we have examined the performance of two nested ILS algorithms using the example application to the quadratic assignment problem. Computational results on large QAP instances have shown that after appropriate tuning, the HILS(2) algorithm tested reaches typically a significantly better performance than the underlying ILS algorithms from which it was derived. Experiments with an hierarchical ILS algorithm that is an example of the third level of the hierarchy, have shown promising results on a class of strongly structured instances that are designed to be hard for local search algorithms.

In summary, we hope that (i) this contributions inspires researchers to consider similar types of hybrid approaches and that (ii) it encourages them to test the hierarchical ILS approach, in particular, for a variety of other applications.

Acknowledgments. This work was supported by the META-X project, an *Action de Recherche Concertée* funded by the Scientific Research Directorate of the French Community of Belgium. Mohamed Saifullah Hussin acknowledges support from the Universiti Malaysia Terengganu. Thomas Stützle acknowledges support from the Belgian F.R.S.-FNRS, of which he is a Research Associate.

References

1. Hoos, H.H., Stützle, T.: Stochastic Local Search—Foundations and Applications. Morgan Kaufmann Publishers, San Francisco (2005)
2. Lourenço, H.R., Martin, O., Stützle, T.: Iterated local search. In: Glover, F., Kochenberger, G. (eds.) Handbook of Metaheuristics, pp. 321–353. Kluwer Academic Publishers, Norwell (2002)
3. Call for Papers: HM2009: 6th International Workshop on Hybrid Metaheuristics (2009), http://www.diegm.uniud.it/hm2009/
4. Smyth, K., Hoos, H.H., Stützle, T.: Iterated robust tabu search for MAX-SAT. In: Xiang, Y., Chaib-draa, B. (eds.) Canadian AI 2003. LNCS (LNAI), vol. 2671, pp. 129–144. Springer, Heidelberg (2003)
5. Cordeau, J.-F., Laporte, G., Pasin, F.: Iterated tabu search for the car sequencing problem. European Journal of Operational Research 191(3), 945–956 (2008)

6. Misevicius, A.: Using iterated tabu search for the traveling salesman problem. Information Technology and Control 32(3), 29–40 (2004)
7. Misevicius, A., Lenkevicius, A., Rubliauskas, D.: Iterated tabu search: an improvement to standard tabu search. Information Technology and Control 35(3), 187–197 (2006)
8. Lozano, M., García-Martínez, C.: An evolutionary ILS-perturbation technique. In: Blesa, M.J., Blum, C., Cotta, C., Fernández, A.J., Gallardo, J.E., Roli, A., Sampels, M. (eds.) HM 2008. LNCS, vol. 5296, pp. 1–15. Springer, Heidelberg (2008)
9. Essafi, I., Mati, Y., Dauzère-Pèréz, S.: A genetic local search algorithm for minimizing total weighted tardiness in the job-shop scheduling problem. Computers & Operations Research 35(8), 2599–2616 (2008)
10. Sahni, S., Gonzalez, T.: P-complete approximation problems. Journal of the ACM 23(3), 555–565 (1976)
11. Burkard, R.E., Çela, E., Pardalos, P.M., Pitsoulis, L.S.: The quadratic assignment problem. In: Pardalos, P.M., Du, D.Z. (eds.) Handbook of Combinatorial Optimization, vol. 2, pp. 241–338. Kluwer Academic Publishers, Dordrecht (1998)
12. Çela, E.: The Quadratic Assignment Problem: Theory and Algorithms. Kluwer Academic Publishers, Dordrecht (1998)
13. Stützle, T.: Iterated local search for the quadratic assignment problem. European Journal of Operational Research 174(1), 1519–1539 (2006)
14. Hansen, P., Mladenović, N.: Variable neighborhood search: Principles and applications. European Journal of Operational Research 130(3), 449–467 (2001)
15. Stützle, T., Fernandes, S.: New benchmark instances for the QAP and the experimental analysis of algorithms. In: Gottlieb, J., Raidl, G.R. (eds.) EvoCOP 2004. LNCS, vol. 3004, pp. 199–209. Springer, Heidelberg (2004)
16. Taillard, É.D.: Comparison of iterative searches for the quadratic assignment problem. Location Science 3(2), 87–105 (1995)
17. Taillard, É.D.: Robust taboo search for the quadratic assignment problem. Parallel Computing 17(4–5), 443–455 (1991)
18. Birattari, M., Stützle, T., Paquete, L., Varrentrapp, K.: A racing algorithm for configuring metaheuristics. In: Langdon, W.B., et al. (eds.) Proceedings of the Genetic and Evolutionary Computation Conference (GECCO-2002), pp. 11–18. Morgan Kaufmann Publishers, San Francisco (2002)
19. Drezner, Z., Hahn, P., Taillard, É.D.: A study of quadratic assignment problem instances that are difficult for meta-heuristic methods. Annals of Operations Research 174, 65–94 (2005)

Incorporating Tabu Search Principles into ACO Algorithms

Franco Arito and Guillermo Leguizamón

Laboratorio de Investigación y Desarrollo en Inteligencia Computacional
Departamento de Informática
Facultad de Ciencias Físico Matemáticas y Naturales
Universidad Nacional de San Luis
San Luis, Argentina
{farito,legui}@unsl.edu.ar
http://www.lidic.unsl.edu.ar

Abstract. ACO algorithms iteratively build solutions to an optimization problem. The solution construction process is guided by pheromone trails which represents a mechanism of adaptation that allows to bias the sampling of new solutions toward promising regions of the search space. Additionally, the bias of the search is influenced by problem dependent heuristic information. In this work we describe an ACO algorithm that incorporates principles of Tabu Search (TS) for the solution construction process. These concepts specifically address the way that TS uses the history of the search to avoid visiting solutions already analyzed. We consider the Quadratic Assignment Problem (QAP) as a case-study, since this problem was also tackled in a closely related research to ours, the one on the usage of external memory in ACO algorithms. The performance of the proposed algorithm is assessed by considering a well-known set of instances of QAP.

1   Introduction

ACO algorithms generate solutions for an optimization problem by a construction mechanism where the selection of the solution component to be added at each step is probabilistically influenced by pheromone trails and (in most of the cases) heuristic information [5].

In this work, we examined the possibility of alternating the way in which the ants select the next solution component by introducing an external memory as *auxiliary mechanism* to make the decisions at each step of the solution construction process. This modification is inspired by Tabu Search (TS). TS explicitly uses the history of the search, both to escape from local minima and to implement an explorative strategy [6]. The adopted approach adds a mechanism with deterministic features to the solution construction step, in opposition with the philosophy of ACO algorithms, which build solutions in a probabilistic way (based on pheromone trails and heuristic information)[1]. The memory structures considered here allow the ants at a specific time, do not make decisions in a probabilistic way, but rather choose solution components in a deterministic way influenced by the values stored in this memory. As this memory will store

---

[1] Another ACO algorithm that introduces deterministic features is Ant Colony System [4].

M.J. Blesa et al. (Eds.): HM 2009, LNCS 5818, pp. 130–140, 2009.

specific information about the search history from the beginning of the algorithm, it allows indeed to focus in not visited regions of the search space (i.e., regions not registered in the memory). In addition, it can concentrates on already visited and promising regions (i.e., registered regions in the memory with values that shows this situation). These memory values allow to apply alternative intensification/diversification mechanisms that, although already present in algorithms ACO, are useful to avoid premature convergence and to improve the algorithm performance respect to optimization problems with specific features. This approach belongs to one of the current trends in ACO algorithms, in which it is incorporated an external memory as a way to seed the solution construction of the ants with a partial solution. [1][2][12][13].

The paper is organized as follows. Section 2 introduces the quadratic assignment problem. In Section 3, we present our proposed memory-based approach. Section 4 gives the results of an extensive computational study. We conclude this work in Section 5 drawing some conclusions and a discussion about future research directions.

2   Quadratic Assignment Problem

The QAP is a $\mathcal{NP}$-hard problem [7] that arises in many real world applications, such as VLSI module placement, scheduling, manufacturing, process communications, statistical data analysis, hospital layout, among others [3]. Also, many combinatorial optimization problems such as the Travelling Salesman Problem (TSP), Maximum Clique Problem (MCP), Graph Partitioning Problem (GPP) can be formulated as QAPs.

The QAP can be described as the problem of assigning a set of objects to a set of locations with given distances between the locations and given flows between the objects. The goal is to place the objects on locations in such a way that the sum of the product between flows and distances is minimal. Given n objects and n locations, two $n \times n$ matrices $A = [a_{ir}]$ and $B = [b_{js}]$, where a_{ir} is the distance between locations i and r, and b_{js} is the flow between objects j y s; the QAP can be formulated as:

$$\min_{\phi \in \Phi} \sum_{i=1}^{n} \sum_{r=1}^{n} a_{\phi_i \phi_r} b_{ir}$$

where $\Phi(n)$ is the set of all possible permutations in $\{1, \ldots, n\}$, and ϕ_j gives the location of object j in the current solution $\phi \in \Phi(n)$.

3   A Memory-Based ACO Algorithm for the QAP

ACO algorithms are among the best performing algorithms for the QAP [11], particularly the $\mathcal{MAX} - \mathcal{MIN}$ Ant System ($\mathcal{MM}$AS-QAP) algorithm [8], [10]. Hence, we have chosen $\mathcal{MM}$AS-QAP as a starting point for our analysis. $\mathcal{MM}$AS-QAP constructs solutions by assigning at each construction step an element to some location. Pheromone trail τ_{ij} indicates to the desirability of assigning element j to location i. $\mathcal{MM}$AS-QAP does not use any heuristic information in the solution construction. The pheromone update is done by lowering the pheromone trails by a constant factor ρ and depositing pheromone on the individual solution components of either the best solution in the current iteration, the best solution found so far by the algorithm, or the best solution found since the last re-initialization of the pheromone trails.

The proposed ACO algorithm differs slightly from the original $\mathcal{MM}$AS-QAP. In the original $\mathcal{MM}$AS-QAP were proposed two ways of constructing solutions, the first one makes at each construction step a probabilistic choice similar to the rule in Ant System [5], whereas a second one uses the pseudo-random proportional action choice rule similar to the rule used in Ant Colony System [4]. Our algorithm makes use of a modification of the second one rule, which is explained in later sections. Also, the pheromone trail reinitialization is not used at all[2]. The proposed algorithm uses local search for improving each candidate solution generated by the ants. Here we use an iterative improvement algorithm (the *2−exchange neighborhood*) where two candidate solutions are neighbors if they differ in the assignment of exactly 2 units to locations, the same that is used in the original $\mathcal{MM}$AS-QAP. The local search algorithm uses a best-improvement pivoting rule.

3.1   Memory Structures

The memory structures in Tabu Search operate by reference to four principal dimensions: recency, frequency, quality, and influence. Our approach makes use of two of them: recency and frequency. Next we explain how they are used in the ACO algorithm.

Recency based memory: This type of memory stores components of the solutions that have changed in the recent past. The usual way of exploiting this type of memory is labeling the selected components of solutions visited recently. What this approach is trying to achieve is to "forbid" certain choices that prevent explore a larger region of the search space or "encourage" certain choices to concentrate on a particular region based on solutions already visited in the recent past. It is important to have in mind that for some instances, a good search process would result in visiting again a previously found solution. Thus, this mechanism aims at continuously stimulating the discovery of new solutions of high quality.

For the QAP, the recency based memory stores the iteration in which the algorithm[3] assigned the object j to location i, for $1 \leq i, j \leq n$. First, this memory allows the ants to make decisions taking into account what objects are the *most recently* assigned to a particular location since they have an associated value near to the current iteration of the algorithm. Second, it allows to make decisions taking into account what objects are the *less recently* assigned to a particular location since they have an associated value far from the current iteration of the algorithm.

The respective ACO algorithm based on recency maintains a $n \times n$ recency based matrix called recency, where recency[i,j] stores the most recent iteration (the last one) in which the object i has been assigned to the location j during the execution of

---

[2] The pheromone trail reinitialization is an additional technique to increase the diversification of the search. We used these technique in previous experiments but the results did not improve the performance of the algorithm, so we decided not to use.

[3] This value is not the iteration value of the algorithm itself, is a value computed according to the number of ants and to the current iteration number.

the algorithm. Therefore, this matrix is used in the process of solution construction in two possible ways:

- Intensification: choose the object that most recently was assigned to the current location (i.e., if the current location is i, it is chosen that object j such that recency[i,j] have the highest iteration number).
- Diversification: choose the object that less recently was assigned to the current location (i.e., if the current location is i, it is chosen that object j such that recency[i,j] have the lowest iteration number).

Frequency based memory: This type of memory stores components of the solutions that are in a solution with more frequency, i.e., it accounts for the number of times that a component is either present in a solution or in a specific position of the solution. The usual way of exploiting this type of memory is labeling the selected components of solutions most frequently chosen. This memory allows to "forbid" that an ant chooses a solution component when it has been frequently chosen in the previous solutions. Thus, the "prohibition" aims at generating solutions that indeed differ of those already generated. In this way, the exploration of the search space is extended. Inversely, this information can be used to "promote" their selection since a most of the ants have chosen them as part of their solutions and therefore they can be considered desirable of being members of a new solution.

For the QAP, the frequency based memory stores the times that object j has been assigned to location i, for $1 \leq i,j \leq n$. Then, this memory allows the ants to make decisions taking into account those elements that were the *most frequently* assigned to a particular location because have associated a high frequency of assignation. On the other hand, it allows to make decisions taking into account those elements that were the *less frequently* assigned to a particular location since they have associated a low frequency assignation.

The respective ACO algorithm maintains a $n \times n$ frequency based matrix called frequency, where the frequency[i,j] stores the quantity of times that the object j has been assigned to the location i during the execution of the algorithm. Then, this matrix is used in the process of solution construction in two possible ways:

- Intensification: choose the object that most times was assigned to the current location (i.e., if the current location is i, it is chosen the object j for which frequency[i,j] is the highest).
- Diversification: choose the object that less times was assigned to the current location (i.e., if the current location is i, it is chosen the object j for which frequency[i,j] is the lowest).

3.2 Solution Construction

When ants construct solutions to the QAP, they assign each object exactly to one location and no location is used by more than one object. The constructed solution corresponds to a permutation $\phi \in \Phi(n)$. The solution construction process involves two steps. In the first step, a location is chosen and then, in the second step an object is assigned to that location. To do that, we randomly choose a location i among those not yet occupied. For the second step we use the pheromone trails, τ_{ij} referring to the desire of assigning the object j in the location i. To assign an object j on an unoccupied location i we used the following rule:

$$j = \begin{cases} T \text{ if } q < q_0 \text{ (Explotation)} \\ R \text{ if } q \geq q_0 \text{ (Exploration)} \end{cases} \tag{1}$$

This rule is similar to the one used by Ant Colony System [5], with a fixed probability q_0 $(0 \leq q_0 \leq 1)$ the ant choose the "best possible element" according to the acquired knowledge (it can be based on the external memory or based on the pheromone trails). With probability $(1 - q_0)$ it is carried out a controlled exploration of new solutions (also in this case, it can be based on the external memory or based on the pheromone trails), where q is a random number uniformly distributed on interval $[0,1]$. T and R are random variables with probability distribution given by Eqs. 2 and 3 respectively as explained in the following.

To promote exploitation, we used the rule:

$$T = \begin{cases} arg\max_{l \in \mathcal{N}_i^k}\{\text{memory}[\text{i},\text{l}]\} \text{ if } r < r_0 \\ arg\max_{l \in \mathcal{N}_i^k}\{\tau_{il}\} \qquad\qquad \text{ if } r \geq r_0 \end{cases} \tag{2}$$

In this rule, with a fixed probability r_0 $(0 \leq r_0 \leq 1)$ is chosen the object that most times was assigned to the current location (memory=frequency) or that most recently was assigned to the current location (memory=recency) whereas with a probability $(1 - r_0)$, the more desirable object is chosen according to the pheromone trails. Variable r is a random number uniformly distributed on the interval $[0,1]$ and where $\mathcal{N}_i^k$ is the set of still unassigned elements for the ant k, that is, those elements that are still to be assigned to location i.

To promote exploration, we used the rule:

$$R = \begin{cases} arg\min_{l \in \mathcal{N}_i^k}\{\text{memory}[\text{i},\text{l}]\} \text{ if } p < p_0 \\ \dfrac{\tau_{ij}(t)}{\sum_{l \in \mathcal{N}_i^k} \tau_{il}(t)} \qquad\qquad \text{ if } p \geq p_0 \end{cases} \tag{3}$$

In this rule, with a fixed probability p_0 $(0 \leq p_0 \leq 1)$ is chosen the element that less times was assigned to the current location (memory=frequency) or the element that less recently was assigned to the current location (memory=recency) whereas with probability $(1 - p_0)$ is chosen an element according to the basic selection rule similar to the rule of the Ant System algorithm (notice that in this case we do not use heuristic information at all). Variable p is a uniformly distributed random number in the interval $[0,1]$.

4   Computational Study

We tested our memory-based ACO algorithm on the QAP instances proposed by Stützle and Fernandes [9] of size 50 (we choose 4 instances from each class). These instances

were generated in such a way that (i) the characteristics of the instances are systematically varied and (ii) they are large enough to allow systematic studies on the dependence
of the performance of metaheuristics on different instance characteristics. The tested instances include:

- **GridRandom:** Grid-based distance matrix and random flows;
- **GridStructured:** Grid-based distance matrix and structured flows;
- **GridStructuredPlus:** Grid-based distance matrix and structured flows with con-
  nections among clusters of objects.
- **RandomRandom:** Random distance matrix and random flows;
- **RandomStructured:** Random distance matrix and structured flows;
- **RandomStructuredPlus:** Random distance matrix and structured flows with con-
  nections among clusters of objects;

All the experimental results are measured across 30 independent trials of the algorithms
and the code was run on a Intel Core 2 Duo 2.13 GHz processor and 1 GB RAM,
running SUSE Linux 10.2. For all instances we ran 1000 iterations. The parameter
values for $\mathcal{MM}$AS-QAP are set as proposed in [10] except that the value of ρ is set
to 0.2 which results in slightly better performance than the setting $\rho = 0.8$ proposed in
the literature and we use $m = 20$ ants (all ants apply local search to the solution they
generate).

We compare the algorithms using the average percentage excess over the best-known
solutions (these were provided to us by the authors of [9]) and for each comparison
we use the one-way ANOVA test to check the statistical significance of the observed
differences in performance (all the distributions were tested to be normal, according to
the Kolmogorov-Smirnov test). Also we used the Tukey's test to find which means are
significantly different from one another.

4.1 Parameters Setting for the Memory-Based $\mathcal{MM}$AS

In this section, we present the results on the parameter settings for the memory-based
$\mathcal{MM}$AS (we considered the four variants of the algorithm according the type of memory used). The next section will be concerned with a comparison of the memory-based
$\mathcal{MM}$AS with $\mathcal{MM}$AS-QAP. In what follows, we refer to the variants of the memory-
based $\mathcal{MM}$AS by using the abbreviations in dependence of the way of combining the
different memories (i.e., the frequency based memory in Eq. 2 or the recency based
memory in Eq. 3). For example, $\mathcal{MM}$AS-fr stands for the variant that uses the frequency based matrix on Eq. 2 and the recency based matrix on Eq. 3. A similar reasoning applies to $\mathcal{MM}$AS-ff, $\mathcal{MM}$AS-rf, and $\mathcal{MM}$AS-rr.

Probability of using external memory in exploitation: We tested the influence of
the parameter r_0 in Eq. 2. The considered values are $\{0.2, 0.5, 0.8\}$, to reflect the fact
that was used respectively with a limited, medium and high frequency. The results in
Table 1 to 4 suggest that a high use of the memory ($r_0 = 0.8$) is more suitable than the
pheromone trails in the intensification process.

Probability of using external memory in exploration: We tested the influence of the
parameter p_0 in Eq. 3. The considered values are $\{0.001, 0.01, 0.1\}$. In this case, the use

Table 1. Comparison of values for parameters p_0 and r_0 in $\mathcal{MM}$AS-ff. The best results for each instance class are indicated in bold-face.

p_0	r_0	Grid Random	Grid Structured	Grid StructuredPlus	Random Random	Random Structured	Random StructuredPlus
0.001	0.2	0.1235	0.1504	0.1668	0.0639	0.0964	0.0729
0.01	0.2	0.1215	0.1652	0.1929	0.0814	0.1002	0.1045
0.1	0.2	0.1530	0.2892	0.3321	0.1480	0.4166	0.4385
0.001	0.5	0.1019	0.1051	0.1242	0.0351	0.0587	0.0591
0.01	0.5	0.1046	0.1279	0.1279	0.0511	0.0615	0.0711
0.1	0.5	0.1535	0.2619	0.2894	0.1475	0.3700	0.3646
0.001	0.8	**0.0679**	**0.0442**	0.0616	0.0390	**0.0208**	**0.0150**
0.01	0.8	0.0729	0.0615	**0.0578**	**0.0360**	0.0371	0.0256
0.1	0.8	0.1498	0.2223	0.2455	0.1186	0.2436	0.2206

Table 2. Comparison of values for parameters p_0 and r_0 in variant $\mathcal{MM}$AS-fr. The best results for each instance class are indicated in bold-face.

p_0	r_0	Grid Random	Grid Structured	Grid StructuredPlus	Random Random	Random Structured	Random StructuredPlus
0.001	0.2	0,1192	0,1520	0,1685	0,0645	0,1162	0,1072
0.01	0.2	0,1147	0,1565	0,1770	0,0745	0,0969	0,1024
0.1	0.2	0,1516	0,2894	0,3170	0,1516	0,4194	0,4169
0.001	0.5	0,0895	0,1001	0,1060	0,0438	0,0998	0,0501
0.01	0.5	0,0976	0,0987	0,1049	0,0572	0,0799	0,0601
0.1	0.5	0,1499	0,2602	0,2770	0,1247	0,3081	0,3186
0.001	0.8	**0,0645**	**0,0564**	0,0765	**0,0280**	**0,0220**	0,0447
0.01	0.8	0,0705	0,0690	**0,0745**	0,0375	0,0642	**0,0218**
0.1	0.8	0,1375	0,2163	0,2437	0,0974	0,2258	0,2307

Table 3. Comparison of values for parameters p_0 and r_0 in variant $\mathcal{MM}$AS-rf. The best results for each instance class are indicated in bold-face.

p_0	r_0	Grid Random	Grid Structured	Grid StructuredPlus	Random Random	Random Structured	Random StructuredPlus
0.001	0.2	0,1279	0,1483	0,1780	0,0533	0,1061	0,0815
0.01	0.2	0,1104	0,1499	0,1816	0,0815	0,1053	0,1235
0.1	0.2	0,1634	0,3168	0,3344	0,1537	0,4213	0,4549
0.001	0.5	0,1102	0,1124	0,1295	0,0434	0,0390	0,0396
0.01	0.5	0,1126	0,1252	0,1440	0,0510	0,0606	0,0400
0.1	0.5	0,1613	0,2796	0,3146	0,1545	0,3951	0,3918
0.001	0.8	**0,0751**	**0,0561**	**0,0707**	0,0423	**0,0222**	**0,0285**
0.01	0.8	0,0818	0,0807	0,0874	**0,0358**	0,0513	0,0361
0.1	0.8	0,1530	0,2685	0,2387	0,1382	0,2721	0,2595

Table 4. Comparison of values for parameters p_0 and r_0 in variant $\mathcal{MM}$AS-rr. The best results for each instance class are indicated in bold-face.

p_0	r_0	Grid Random	Grid Structured	Grid StructuredPlus	Random Random	Random Structured	Random StructuredPlus
0.001	0.2	0,1269	0,1596	0,1791	0,0602	0,1112	0,1021
0.01	0.2	0,1268	0,1655	0,1945	0,0694	0,1226	0,0929
0.1	0.2	0,1570	0,2902	0,3248	0,1511	0,4140	0,3836
0.001	0.5	0,0979	0,1119	0,1155	0,0462	0,0600	0,0662
0.01	0.5	0,1106	0,1331	0,1527	0,0524	0,0603	0,0604
0.1	0.5	0,1532	0,2730	0,2666	0,1371	0,3241	0,3323
0.001	0.8	**0,0716**	**0,0610**	0,0708	**0,0417**	**0,0307**	0,0303
0.01	0.8	0,0750	0,0772	**0,0637**	0,0451	0,0686	**0,0228**
0.1	0.8	0,1490	0,2577	0,2415	0,1211	0,2215	0,2658

Table 5. Comparison of the best-found values from $\mathcal{MM}$AS-ff, $\mathcal{MM}$AS-fr, $\mathcal{MM}$AS-rf, and $\mathcal{MM}$AS-rr plus the respective p-values obtained from a one-way ANOVA test

Algorithm	Grid Random	Grid Structured	Grid StructuredPlus	Random Random	Random Structured	Random StructuredPlus
$\mathcal{MM}$AS-ff	0.0679	0.0442	0.0616	0.0390	0.0208	0.0150
$\mathcal{MM}$AS-fr	0,0645	0,0564	0,0765	0,0280	0,0220	0,0447
$\mathcal{MM}$AS-rf	0,0751	0,0561	0,0707	0,0423	0,0222	0,0285
$\mathcal{MM}$AS-rr	0,0716	0,0610	0,0708	0,0417	0,0307	0,0303
p-value	**0,000027**	**0.0425**	**0.0051**	0.5816	0.7236	0.8605

Table 6. Comparison of the best-found values from $\mathcal{MM}$AS-ff, $\mathcal{MM}$AS-fr, $\mathcal{MM}$AS-rf, $\mathcal{MM}$AS-rr and $\mathcal{MM}$AS plus the respective p-values obtained from a one-way ANOVA test

Algorithm	Grid Random	Grid Structured	Grid StructuredPlus	Random Random	Random Structured	Random StructuredPlus
$\mathcal{MM}$AS-ff	0.0679	0.0442	0.0616	0.0390	0.0208	0.0150
$\mathcal{MM}$AS-fr	0,0645	0,0564	0,0765	0,0280	0,0220	0,0447
$\mathcal{MM}$AS-rf	0,0751	0,0561	0,0707	0,0423	0,0222	0,0285
$\mathcal{MM}$AS-rr	0,0716	0,0610	0,0708	0,0417	0,0307	0,0303
$\mathcal{MM}$AS	0,1286	0,1734	0,1944	0,0740	0,1025	0,1024
p-value	≈ 0	≈ 0	≈ 0	≈ 0	≈ 0	≈ 0

of memory should be more restricted because the pheromone trails are the main component in ACO algorithms. This was confirmed in the results in Table 1 to Table 4, since the best performance was obtained, in almost the cases, with the smallest probability.

In Table 5 we compared the best results from each of the variants, in almost all cases the best variant was the setting that had $p_0 = 0.001$ and $r_0 = 0.8$. Although the respective mean values are not showed in Table 5, we analize them by considering the one-way ANOVA test and we can see that the statistical differences are significant

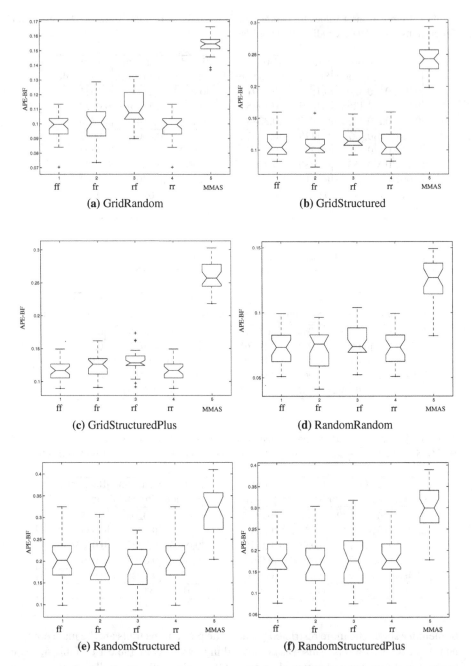

Fig. 1. Boxplots of the one-way ANOVA test across the six different instances classes. The y-axis (APE-BF) gives the average percentage excess from Best Known solution. On the x-axis are shown the respective variants: ff ($\mathcal{MM}$AS-ff), fr ($\mathcal{MM}$AS-fr), rf ($\mathcal{MM}$AS-rf), rr ($\mathcal{MM}$AS-rr), and MMAS ($\mathcal{MM}$AS-QAP).

for the algorithm $\mathcal{MM}$AS-rf respect to the other tree variants, in particular for tree instances class (GridRandom, GridStructured and GridStructuredPlus) as can be observed in the last row in Table 5. Also, according to the Tukey method the means of algorithm $\mathcal{MM}$AS-rf are significantly different to those of the algorithms $\mathcal{MM}$AS-ff, $\mathcal{MM}$AS-fr and $\mathcal{MM}$AS-rr.

4.2 Comparison of $\mathcal{MM}$AS-QAP and the Memory-Based $\mathcal{MM}$AS

In this section we compare the performance of the original $\mathcal{MM}$AS-QAP to the best performing variants mentioned in the previous section. The results in Table 6 shows that the original $\mathcal{MM}$AS-QAP achieved a lower performance than all the four variants of the memory-based $\mathcal{MM}$AS on the set of instances considered in our study. The above assessment is supported considering the p-values (see last row in Table 6) which all of them are near to zero after including the respective values from $\mathcal{MM}$AS-QAP in the one-way ANOVA test with respect to the four variants of the memory-based $\mathcal{MM}$AS. Additionally, Fig. 1 displays a set of boxplots where the respective differences between the algorithms can be observed. Finally, we remark that during the experiments we were able to improve the Best Known solution of the instance:

– RandomStructuredPlus.974824391.n50.K10.m10.A100.00.B1.00.sp10.00.dat

the last best know solution was 7646 and the new one is 7500, this was obtained by the algorithm $\mathcal{MM}$AS-rf. The Best Known solution is:

$$\phi = (38, 20, 19, 14, 10, 49, 36, 9, 11, 3, 22, 1, 27, 24, 47, 46, 39, 15, 13, 7, 44,$$
$$34, 31, 2, 32, 45, 16, 23, 21, 17, 40, 8, 42, 6, 18, 12, 28, 4, 5, 29, 41, 25, 37,$$
$$35, 26, 48, 30, 0, 43, 33)$$

5   Conclusions and Future Work

In this paper, we presented a variant of the algorithm $\mathcal{MM}$AS-QAP that incorporates Tabu Search principles. This is achieved trough the use of an external memory, that stores information collected during the search. The external memory complements the information contained in the pheromone trails, giving specific information about the search history since the beginning of the algorithm that is useful during the solution construction process. The computational results of this adaptation show that the introduction of the external memory can improve the performance, properly setting the parameters involved with the use of this memory, of a variant of the original $\mathcal{MM}$AS-QAP.

The results obtained agree with the positive results reported in similar approaches [1], [2], [12].

As future work we plan to conduct a more detailed study of the parameters settings involved with the use of the external memory that includes Space Filling analysis (Latin Hypercube Design). Also we plan to adapt the idea of the memory into another ACO algorithms, and we consider the possibility of adapting the idea of external memory into ACO algorithms applied to other combinatorial optimization problems.

Acknowledgments. The authors wish to thank Prof. Thomas Stützle by the source code of ACOTSP Version 1.0. available on-line[4] in which it is based our algorithm.

References

1. Acan, A.: An external memory implementation in ant colony optimization. In: Dorigo, M., Birattari, M., Blum, C., Gambardella, L.M., Mondada, F., Stützle, T. (eds.) ANTS 2004. LNCS, vol. 3172, pp. 73–84. Springer, Heidelberg (2004)
2. Acan, A.: An external partial permutations memory for ant colony optimization. In: Raidl, G.R., Gottlieb, J. (eds.) EvoCOP 2005. LNCS, vol. 3448, pp. 1–11. Springer, Heidelberg (2005)
3. Cela, E.: The Quadratic Assignment Problem: Theory and Algorithms. Kluwer Academic Publishers, Dordrecht (1998)
4. Dorigo, M., Gambardella, L.M.: Ant Colony System: A cooperative learning approach to the traveling salesman problem. IEEE Transactions on Evolutionary Computation 1(1), 53–66 (1997)
5. Dorigo, M., Stützle, T.: Ant Colony Optimization. MIT Press, Cambridge (2004)
6. Glover, F., Laguna, M.: Tabu Search. Kluwer Academic Publishers, Norwell (1997)
7. Sahni, S., Gonzalez, T.: P-complete approximation problems. Journal of the ACM 23(3), 555–565 (1976)
8. Stützle, T.: $\mathcal{MAX} - \mathcal{MIN}$ for the quadratic assignment problem. Technical report, AIDA-97-4, FG Intellektik, FB Informatik, TU Darmstadt, Germany (1997)
9. Stützle, T., Fernandes, S.: New benchmark instances for the QAP and the experimental analysis of algorithms. In: Gottlieb, J., Raidl, G.R. (eds.) EvoCOP 2004. LNCS, vol. 3004, pp. 199–209. Springer, Heidelberg (2004)
10. Stützle, T., Hoos, H.: $\mathcal{MAX} - \mathcal{MIN}$ ant system. Future Generation Computer Systems 16(8), 889–914 (2000)
11. Stützle, T., Dorigo, M.: Aco algorithms for the quadratic assignment problem. In: Corne, D., Dorigo, M., Glover, F. (eds.) New Ideas in Optimization, London, UK, pp. 33–50. McGraw-Hill, New York (1999)
12. Tsutsui, S.: cAS: Ant colony optimization with cunning ants. In: Runarsson, T.P., Beyer, H.-G., Burke, E.K., Merelo-Guervós, J.J., Whitley, L.D., Yao, X. (eds.) PPSN 2006. LNCS, vol. 4193, pp. 162–171. Springer, Heidelberg (2006)
13. Wiesemann, W., Stützle, T.: Iterated ants: An experimental study for the quadratic assignment problem. In: Dorigo, M., Gambardella, L.M., Birattari, M., Martinoli, A., Poli, R., Stützle, T. (eds.) ANTS 2006. LNCS, vol. 4150, pp. 179–190. Springer, Heidelberg (2006)

---

[4] http://www.aco-metaheuristic.org/aco-code

A Hybrid Solver for Large Neighborhood Search: Mixing Gecode and EasyLocal++

Raffaele Cipriano[1], Luca Di Gaspero[2], and Agostino Dovier[1]

[1] DIMI
{cipriano,dovier}@dimi.uniud.it
[2] DIEGM Università di Udine, via delle Scienze 208, I-33100, Udine, Italy
l.digaspero@uniud.it

Abstract. We present a hybrid solver (called GELATO) that exploits
the potentiality of a Constraint Programming (CP) environment (Gecode)
and of a Local Search (LS) framework (EasyLocal++). GELATO allows
to easily develop and use hybrid meta-heuristic combining CP and LS
phases (in particular Large Neighborhood Search). We tested some
hybrid algorithms on different instances of the Asymmetric Traveling
Salesman Problem: even if only naive LS strategies have been used,
our meta-heuristics improve the standard CP search, in terms of both
goodness of the solution reached and execution time. GELATO will
be integrated into a more general tool to solve Constraint Satisfac-
tion/Optimization Problems. Moreover, it can be seen as a new library
for approximate and efficient searching in Gecode.

1 Introduction

Combinatorial problems like planning, scheduling, timetabling and, in general,
resource management problems, are daily handled by industries, companies, hospitals and universities. Their intractability however poses challenging problems
to the programmer: ad-hoc heuristics that are adequate for one problem are often useless for others [16]; classical techniques like integer linear programming
(ILP) need a big tuning in order to be effective; any change in the specification of the problem requires restarting almost from scratch. In the last years
many efforts have been put in the development of general techniques that allow
high-level primitives to encode search heuristics. Noticeable examples of these
techniques are constraint programming (CP—with the various labeling heuristics) [12] and local search (LS—with the various techniques to choose and visit
the neighborhood). We are not dealing with the problem of finding the optimum
of a problem but with a *reasonable* solution to be computed in reasonable time.
The future seems to stay in the combinations of these techniques in order to
exploit the best aspect of each technique for the problem at hand (after a tuning
of the various parameters on small instances of the problem considered). This
is also witnessed by the success of the CP-AI-OR meetings [15]. In particular,
Large Neighborhood Search (LNS) can be viewed as a particular heuristic for
local search that strongly relies on a constraint solver [3] and it is a reasonable
way to blend the inference capabilities of LS and CP techniques.

M.J. Blesa et al. (Eds.): HM 2009, LNCS 5818, pp. 141–155, 2009.

In this work we develop a general framework called GELATO that integrates CP and LS techniques, defining a general LNS meta-heuristic that can be modified, tuned and adapted to any optimization problem, with a limited programming effort. Instead of building a new solver from scratch, we based our framework on two state-of-the-art, existing systems: the Gecode CP environment [13] and the LS framework EasyLocal[++] [5]. We choose these two systems (among other ones, such as [11,10,7]) because both of them are free and open, strong and complete, written in C^{++}, and with a growing community using them.

We show how to model a LNS algorithm combining CP and LS in GELATO: this modeling is easy to implement and it allows efficient computations. We test the framework on hard Asymmetric Travel Salesman Problem instances and show its effectiveness w.r.t. the traditional use of a pure CP approach.

The results of this paper will be combined to those of [2] so as to obtain a multi-language system able to model and solve combinatorial problems. This tool will comprise three main parts: a modeling component, a translator, and a solver. In the modeling component the user will be able to define in a high-level style the problem and the instance he/she wants to solve and the algorithm to use (CP search, possibly interleaved with LS, integer linear programming, heuristics or meta-heuristics phases). The translation component will handle the compilation of the model and the meta-algorithm defined by the user into the solver frameworks, like Gecode or others. In the third phase, the overall compiled program will be run on the problem instance specified by the user and the various solvers will interact as set by the user in the model. A side-effect of our work is a new library of search strategies for the Gecode system. Gecode is a free development environment for CP; GELATO provides some primitives that allow a Gecode user to easily develop approximate search algorithms, based on LS (in particular, with LNS). On the other hand, GELATO can be considered also as a plug-in for the EasyLocal[++] system, a framework that allows the user to easily develop, test and combine several Local Search algorithms. With GELATO, an EasyLocal[++] user can easily exploit the benefits of CP in its LS algorithms.

2   Preliminary Concepts

Discrete or continuous problems can be formalized using the concept of *Constraint Satisfaction Problem* (CSP). A CSP is defined as the problem of associating values (taken from a set of domains) to variables subject to a set of constraints. A *solution* of a CSP is an assignment of values to all the variables so that all the constraints are satisfied. In some cases not all solutions are equally preferable, but we can associate a cost function to the variable assignments. In these cases we talk about *Constraint Optimization Problems* (COPs), and we are looking for a solution that minimizes the cost value. The solution methods for CSPs and COPs can be split into two categories:

- *Complete methods*, which systematically explore the whole solution space in search of a feasible (for CSPs) or an optimal (for COPs) solution.

– *Incomplete methods*, which rely on heuristics that focus only on some areas of the search space to find a feasible solution (CSPs) or a "good" one (COPs).

2.1 Constraint Programming Basics

Constraint Programming (CP) [12] is a declarative programming methodology parametric on the constraint domain. Combinatorial problems are usually encoded using constraints over *finite domains*, currently supported by all CP systems (e.g., [13,10,11]).

A CSP $\mathcal{P}$ is modelled as follows: a set $X = \{x_1, \ldots, x_k\}$ of *variables*; a set $D = \{D_1, \ldots, D_k\}$ of *domains* associated to the variables (i.e., if $x_i = d_i$ then $d_i \in D_i$); a set $\mathcal{C}$ of *constraints* (i.e., relations) over $dom = D_1 \times \cdots \times D_k$. $\langle d_1, \ldots, d_k \rangle \in dom$ satisfies a constraint $C \in \mathcal{C}$ iff $\langle d_1, \ldots, d_k \rangle \in C$. A tuple $d = \langle d_1, \ldots, d_k \rangle \in dom$ is a solution of a CSP $\mathcal{P}$ if d satisfies every constraint $C \in \mathcal{C}$. The set of solutions of $\mathcal{P}$ is denoted with $sol(\mathcal{P})$. If $sol(\mathcal{P}) \neq \emptyset$, then $\mathcal{P}$ is consistent. Often a CSP is associated to a function $f : sol(\mathcal{P}) \to E$ where $\langle E, \leq \rangle$ is a well-ordered set (e.g., $E = \mathbb{N}$ or $E = \mathbb{R}$). A COP is a CSP with an associated function f. A solution for this COP is a solution $d \in sol(\mathcal{P})$ that minimizes the function f: $\forall e \in sol(\mathcal{P})(f(d) \leq f(e))$.

This paradigm is usually based on *complete methods* that analyze the search space alternating deterministic phases (constraint propagation—values that cannot be assigned to any solution are removed by domains) and non-deterministic phases (variable assignment—a variable is selected and a value from its domain is assigned to it). This process is iterated until a solution is found or unsatisfiability is reached; in the last case the process backtracks to the last choice point (i.e., the last variable assignment) and tries other assignments.

2.2 Local Search Basics

Local Search (LS) methods (see [1,4]) are a family of meta-heuristics to solve CSPs and COPs, based on the definition of *proximity* (or *neighborhood*): a LS algorithm typically moves from a solution to a near one, trying to improve an objective function, iterating this precess. LS algorithms generally focus the search only in specific areas of the search space, so they are *incomplete methods*, in the sense that they do not guarantee to find a feasible (or optimal) solution, but they search non-systematically until a specific stop criterion is satisfied.

To define a LS algorithm for a given COP, three parameters must be defined: the *search space*, the *neighborhood relation*, and the *cost function*. Given a COP P, we associate a *search space* S to it, so that each element $s \in S$ represents a solution of P. An element s is a *feasible* solution iff it fulfills the constraints of P. S must contain at least one feasible solution.

For each element $s \in S$, a set $\mathcal{N}(s) \subseteq S$ is defined. The set $\mathcal{N}(s)$ is called the *neighborhood* of s and each member $s' \in \mathcal{N}(s)$ is called a *neighbor* of s. In general $\mathcal{N}(s)$ is implicitly defined by referring to a set of possible *moves*, which define transitions between solutions. Moves are usually defined in an intensional fashion, as local modifications of some part of s. A *cost function* f, which associates to each element $s \in S$ a value $f(s) \in E$, assesses the quality of the solution.

f is used to drive the search toward good solutions and to select the move to perform at each step of the search. For CSP problems, the cost function f is generally based on the so-called *distance to feasibility*, which accounts for the number of constraints that are violated. A LS *algorithm* starts from an initial solution $s_0 \in S$, and uses the moves associated with the neighborhood definition to navigate the search space: at each step it makes a transition between one solution s to one of its neighbors s', and this process is iterated. When the algorithm makes the transition from s to s', we say that the corresponding move m has been accepted. The selection of moves is based on the values of the cost function and it depends on the specific LS technique.

Large Neighborhood Search (LNS) is a LS method that relies on a particular definition of the neighborhood relation and of the strategy to explore the neighborhood. Differently from traditional LS methods, an existing solution is not modified just by making small changes to a limited number of variables (as is typical with LS move operators), instead a subset of the problem is selected and searched for improving solutions. The subset of the problem can be represented by a set FV of variables, that we call *free variables*, which is a subset of the variables X of the problem. Defining FV corresponds to define a neighborhood relation.

For example, a LS move could be to swap the values of two variables, or, more generally the permutation of the values of a set of variables. Another possibility for a neighborhood definition is to keep the values of some variables and to leave the other variables totally free, constrained only by their domains.

Three aspects are crucial in LNS definition, w.r.t. the performance of this technique: *(1) which* and *(2) how many variables* have to be selected (i.e., the definition of FV), and *(3) how to perform the exploration* on these variables. Let us briefly analyze these three key-points, starting from the third one.

(3) Given FV, the *exploration* can be performed with any searching technique: CP, Operation Research algorithms, and so on. We can be interested in searching for: the best neighborhood; the best neighborhood within a certain exploration timeout; the first improving neighborhood; the first neighborhood improving the objective function of at least a given value, and so on. (2) Deciding *how many variables* will be free ($|FV|$) affects the time spent on every large neighborhood exploration and the improvement of the objective function for each exploration. A small FV will lead to very efficient and fast search, but with very little improvement of the objective function. Otherwise, a big FV can lead to big improvement at each step, but every single exploration can take a lot of time. This trade-off should be investigated experimentally, looking at a dimension of FV that leads to fast enough explorations and to good improvements. Obviously, the choice of $|FV|$ is strictly related to the search technique chosen (e.g., a strong technique can manage more variables than a naive one) and to the use or not of a timeout. (1) The choice of *which variables* will be included in FV is strictly related to the problem we are solving: for simple and not too structured problems we can select the variables in a naive way (randomly, or iterating between given sets of them); for complex and well-structured problems, we should define FV cleverly, selecting the variables which are most likely to give an improvement to the solution.

2.3   Hybridization of CP and LS

Two major types of approaches to combine the abilities of CP and LS are presented in the literature [6,8]:

1. a systematic-search algorithm based on constraint programming can be improved by inserting a LS algorithm at some point of the search procedure, e.g.: (a) at a leaf (i.e., on complete assignments) or on an internal node (i.e., on a partial assignment) of the search tree explored by the constraint programming procedure, in order to improve the solution found; (b) at a node of the search tree, to restrict the list of child-nodes to explore; (c) to generate in a greedy way a path in the search tree;
2. a LS algorithm can benefit of the support of constraint programming, e.g.: (a) to analyze the neighborhood and discarding the neighboring solutions that do not satisfy the constraints; (b) to explore a fragment of the neighborhood of the current solution; (c) to define the search of the best neighboring solution as a problem of constrained optimization (COP).

In these hybrid methods one of the two paradigms is the master and the second one acts as a slave, supporting the master at some point of the search algorithm. Paradigms not based on the master-slave philosophy have also been proposed. In [9] LS and constraint propagation are split in their components, allowing the user to manage the different basic operators (neighborhood exploration, constraint propagation and variable assignment) at the same level. In [7] CP and LS are combined in a programming language (COMET), that supports both modeling and search abstractions, and where constraint programming is used to describe and control LS.

LNS is a LS strategy that can be naturally implemented using CP, leading to hybrid algorithms that involves the approaches 2(a), 2(b) and 2(c) of the above enumeration. CP can manage and exhaustively explore a lot of variables subjected to constraints, so it is a perfect method for the exploration of large neighborhoods.

2.4   The Solvers Used

There are several tools commonly used by the community to solve CSPs and COPs with the CP and LS paradigms. We focused on Gecode , for the CP aspects, and EasyLocal^{++}, for the LS algorithms, for three main reasons: goodness of the solvers (they both are very strong, complete and efficient); guarantee of maintenance during time (they have a growing user community and they are actively maintained by their developers); ease of integration (they are free, open and written in C++, so they can be employed as C++ libraries). Moreover, all these characteristics ensure the possibility in the future of easily integrating these solvers with other C++ libraries. Here we can give only a short explanation of these two solvers, suggesting the reader to take a look at [13] for Gecode and [5] for EasyLocal^{++}.

Gecode. It is an open, free, portable, accessible, and efficient environment for developing constraint-based systems and applications. It is implemented in C++

and offers competitive performances w.r.t. both runtime and memory usage. It implements a lot of data structures, constraints definitions, and search strategies, allowing also the user to define his own ones.

In the Geode philosophy, a model is implemented using *spaces*. A space is the repository for variables, constraints, objective function, searching options. Being C++ an object-oriented language, the modeling approach of Gecode exploits the *inheritance*: a model must implement the class *Space*, and the subclass constructor implements the actual model. In addition to the constructor, a model must implement some other functions(e.g., performing a copy of the space, returning the objective function, ...).

A Gecode space can be asked to perform the propagation of the constraints, to find the first solution (or the next one) exploring the search tree, to find the best solution in the whole search space.

EasyLocal++. It is an object-oriented framework that allows to design, implement and test LS algorithms in an easy, fast and flexible way. The basic idea of EasyLocal++(briefly EL) is to capture the essential features of most LS metaheuristics, and their possible compositions. This allows the user to address the design and implementation issues of new LS heuristics in a more principled way.

The frameworks are characterized by the inverse control mechanism: the functions of the framework call the user-defined ones and not the other way round. The framework thus provides the full control structures for the invariant part of the algorithms, and the user only supplies the problem specific details.

Modeling a problem using EL means to define the C++ classes representing the basic concepts of a LS algorithm: e.g., the structure of a solution (or *State*, in the EL language), the neighborhood (or *Move*), the cost function, the strategy to navigate the neighborhoods (*NeighborhoodExplorer*). Once these basic concepts are defined (in C++ classes that inherit from EL base classes and implement their characteristics), the user selects the desired LS algorithm and run it. EL provides a wide range of LS heuristic and meta-heuristic (Hill Climbing, Steepest Descent, Tabu Search, Multi Neighborhood, ...), allowing also the development of new ones.

3   A Hybrid Solver for Large Neighborhood Search

In this section we describe the hybrid solver we have developed to implement LNS algorithms that exploit a CP solver in the exploration of the large neighborhoods; we called it GELATO (Gecode+EasyLocal = A Tool for Optimization). We first define the basic elements of this tool, in a general way, without any implementation detail. Then we explain the choices we made to integrate these tools in a unique framework.

3.1   The Key Ingredients of the Solver

GELATO is made-up of five main elements (see Fig. 1): a *constraint model*, a *move enumerator*, a *move performer*, a *CP solver*, and a *LS framework*. The constraint model specify the problem we want to solve, according with the definition

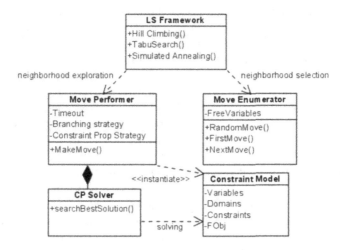

Fig. 1. A UML class diagram of the main components of GELATO

given in Section 2.1. The move enumerator and the move performer define the LNS characteristics (definition of the neighborhood and how to explore it, as in 2.2). The CP solver takes care of exploring the neighborhood. The LS framework manages all these element together into a LS algorithm. We briefly analyze each element.

Constraint model. Here we define the variables, the domains, the constraints and objective function of the problem we want to solve. At these level we do not specify neither the instance input characteristics, that will be passed to the model as a parameter at run time, nor the search options (variable and value selection, timeout, ...), that will be specified by the *move performer* when the model will be actually solved. In GELATO, the constraint model is specified using the Gecode language, but in general it can be expressed using any high-level modeling language (SICStus Prolog, Minizinc, OPL) or low level language, according to the CP solver that will be used. This model will be actually instantiated and solved by the CP solver during the exploration of a large neighborhood. It can be also used to find good starting solution for the main LS algorithm (hybrid approach 1(a) in the enumeration of section 2.3).

Move enumerator (m.e.). It deals with the definition of the set FV for the LNS, specifying which variable of the constraint model will be free in the exploration of the neighborhood. According with the *LS framework*, we can specify several kind of move enumerator: e.g., a random m.e., that randomly selects a specified number of variables of the problems; an iterative m.e., that iterates among all the N combination on the variables of the problem; a sliding m.e., that considers sliding window of K variables (i.e., $FV = \{X_1, \ldots, X_{1+k}\}$, then $FV = \{X_2, \ldots, X_{2+k}\}$ and so on).

Move performer (m.p.). A move performer collects all the information about how to search a large neighborhood, like searching timeout, variable and value selection, constraint propagation policy, branching strategies and so on. It instantiates the *constraint model*, according to the definition of the problem, the instance, the move (given by the m.e.), and the CP solver. It invokes the CP solver passing all these information and obtain the result of the exploration.

CP solver. It is attached to a m.p. and must be able to recognize the *constraint model* specified and to perform an exploration on the neighborhoods, with the searching parameters given by them m.p.. It is expected to return the best neighborhood found (according to the m.p. policy) and the value of the objective function for it. In GELATO the *CP solver* used is Gecode, but in principle, there is no restriction about what kind of solver to use: a solver that can be easily interfaced with the other components could be a good choice.

LS framework. It defines the high level interaction between the above components, building up a real LS algorithm: it navigates the searching space through the neighborhoods defined by the m.e., exploring them using the m.p.. Any traditional LS search algorithm can be used at this step (e.g., Hill Climbing, Steepest Descent, Tabu Search). Also meta-heuristic LS algorithm are allowed (e.g., Multi-Neighborhood, Token Ring Search), if different kinds of m.e. and m.p. are defined for the same problem. The *LS framework* used in GELATO is EasyLocal++.

3.2   Mixing Gecode and EasyLocal++

Gecode and EasyLocal++ are two frameworks that differs in many ways: they have different aims, use different data structures, and of course implement different algorithms. However, they have some architectural similarities: they both are written in C++; they both have base classes for the basic concepts and ask the user to define his problem implementing these base classes; they provide general algorithms (for CP and LS) that during executions call and use the derived classes developed by the user.

To define a hybrid tool for LNS, the natural choice is to use the EL framework as a main algorithm and Gecode as a module of EL that can be invoked to perform large neighborhood explorations.

According to the architecture of Gecode and EL and to the main components of the tools described in section 3.1, we have defined a set of classes (described below) that make possible the integration between the two frameworks. Only the first two (ELState and GEModel) must be implemented by the user (with derived class), because they contain the problem specific information; the other ones are provided by our tools and the user has only to invoke them (but he is also allowed to define his own ones).

ELState. It is an EL state, that derives from the base class State of the EL framework. It typically contains a vector of variables that represents a solution, and an integer that contains the value of the objective function of the state.

GEModel. It is the Gecode model, and it implements the methods requested to work with the Gecode language and the EL framework. It inherits methods and properties from the Gecode Space class.

LNSMove. Base abstract class for the LNS moves of our framework. An actual move must implement the method `bool containsVar(Var)` that says if a given Var is free w.r.t. the move. We have defined two actual derived classes: District-Move and NDifferentMove. DistrictMove is made up by several sets of variables and of an active set: at a given time t only the variables of the active set are free; the active set can be changed by the move enumerator. NDifferentMove is made up by a single set of variables and a fixed integer N: at a given time t only N variables of the set are free (they can be chosen randomly or iterating between the variables in the set).

MoveEnumerator. Base abstract class for m.e., that cycles between a particular kind of LNSMove. MoveEnumerator actual classes must implement the functions `RandomMove()`, `FirstMove()` and `NextMove()`. `RandomMove()`, given a LNSMove, selects randomly the set of free variables, according to the specific LNSMove definition (e.g., selecting randomly an active set of a DistrictMove). `FirstMove()` and `NextMove()` are used to cycle on all the possible moves (e.g., cycling on all the active neighborhoods of a DistrictMove, or cycling on all the N-combinations of the variables contained in a NDifferentMove). Our tool provides an actual MoveEnumerator for each LNSMove: a DistrictMoveME and a NDifferentME.

MovePerformer. This module apply a given move to a given state, returning the new state reached. We define the GecodeMovePerformer, that takes a ELState and a LNSMove and, according to the defined GEModel class, builds up a Gecode Space and solves it.

LNSNeighborhoodExplorer. It is the main class that EL uses for neighborhood explorations. It wraps together a MoveEnumerator and a MovePerformer.

LNSStateManager. It provides some basics functionalities of an ELState, such as calculating the objective function of a state, and finding an initial state for the search. This last task is performed using CP: an instance of the GEModel of the problem is instantiated and explored until a first feasible solution is reached.

4 A Simple Experiment

We tested GELATO on instances of growing sizes of the Asymmetric Travel Salesman Problem, taken from the TSPLib [14]. In this section we describe the problem, the solving algorithms we used, the experiment we have performed, and the results obtained.

The *Asymmetric Travel Salesman Problem (ATSP)* is defined as follows: given a complete directed graph $G = (V, E)$ and a function c that assigns a cost to each directed edge (i, j), find a roundtrip of minimal total cost visiting each node exactly once. We speak of asymmetric TSP if there exists (i, j) such that $c(i, j) \neq c(j, i)$ (imagine a climbing road).

4.1 CP Model, LNS Definition and LS Algorithm for ATSP

The CP Model we used is chosen from the set of examples in the Gecode package. We used an existing model in order to point out the possibility of using GELATO starting from existing CP models, adding only some little modifications to them (e.g., changing the class headings and a couple of statements in the constructor).

The first state is obtained by a CP search over the same model, without any pre-assigned value of the variables, and without a timeout (because we need an initial feasible solution to start the LS algorithm).

The Large Neighborhood definition we have chosen is the following: given a number $N < |V|$ and given a solution (a tour on the nodes, i.e. a permutation of $|V|$), N variables are randomly selected and left free, while the other ones remain fixed. Therefore, the exploration of the neighborhood consists in exploring the N free variables and it is regulated by the following four parameters: (1) the variables with the smallest domain are selected, (2) the values are chosen randomly, (3) a timeout is set, and (4) the best solution found before reaching the timeout is returned.

The LS algorithm used is a traditional Hill Climbing, with a parameter K. At each step it selects randomly a set of N variables, frees them and searches on them for the best neighborhood, until the timeout expires. If the best neighborhood found is better or equal than the old one, it becomes the new current solution; otherwise (i.e., the value is worse), another random neighborhood is selected and explored (this is said *idle iteration*). The algorithm stops when K consecutive idle iterations have been performed (i.e., stagnation of the algorithm has been detected). Every single large neighborhood is explored using CP.

4.2   Experiments

The experiments have been performed on the following instances, taken from the TSPLib [14]: br17 (that we call *instance 0*, with $|V| = 17$), ftv33 (*instance 1*, $|V| = 34$), ftv55 (*instance 2*, $|V| = 56$), ftv70 (*instance 3*, $|V| = 71$), kro124p (*instance 4*, $|V| = 100$), and ftv170 (*instance 5*, $|V| = 171$).

The instances has been solved either using a pure constraint programming approach in **Gecode** or using different LNS approaches encoded in **GELATO**. The LNS approaches differs for the number of variables of the neighborhood ($|FV|$): 5,10,15 for instance 0; 10,20,30 for instance 1; 10,20,30,40 for instances 2 and 3; 10,20,30,40,50 for instances 4 and 5. We tested different timeouts for the exploration of a single large neighborhood: one is $0.5N$ sec, the other is $0.25N$ sec. For instance 4 and 5 we also tested shorter timeouts. The number K of consecutive idle iteration before stop has been fixed to 50.

The pure **Gecode** CP uses the following parameters: (1) the leftmost variable is selected, and (2) the values are chosen increasingly. We run it only once, since it is deterministic, with a timeout of one hour. We also tried a first-fail strategy (the variable with the smallest domain is selected), but its performance are slightly worse than the leftmost one. This is probably due to the regular structure of the problem.

Since LNS computations uses randomness, we repeated each of them 20 times. During each run, we stored the values of the objective function, that corresponds to improvements of the current solution, together with the running time spent. These data have been aggregated in order to analyze the average behavior of the different LNS strategies on each group of 20 runs. To this aim we perform a discretization of the data on regular time intervals (with a step of 0.01s for instance 0 and 1s for the others); subsequently, for each discrete interval we compute the average value of the objective function on the 20 runs.

4.3   Results

Here we collect the results obtained in the experiments. Every picture shows the comparison of different algorithms on a single instance: in Fig. 2, the pure CP approach is compared with LNS with a timeout of $0.5|FV|s$ for the exploration of a single large neighborhood; in Fig. 3, the CP approach remains the same, but the timeout for LNS is set to $0.25|FV|s$. In Fig. 4 we concentrate on larger instances (4 and 5), analyzing three different timeouts for the exploration of a single neighborhood; we also considered a LNS with $|FV| = 50$, which has been experimentally proved to be effective for instance 5. We also tried neighborhoods with $|FV| = 60$ but they turned out to be computationally intractable (with too many variables the explorations of the neighborhood is not effective).

4.4   Discussion

Let us add some consideration about the results obtained, that can be summarized as: as instances grow, GELATO definitely outperforms Gecode (if we are happy with an approximate solution).

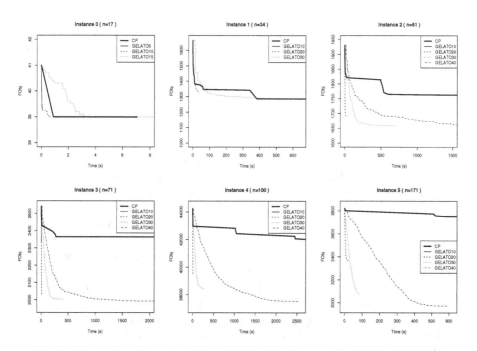

Fig. 2. Comparison between CP and GELATO (timeout for LNS is $0.5|FV|s$)

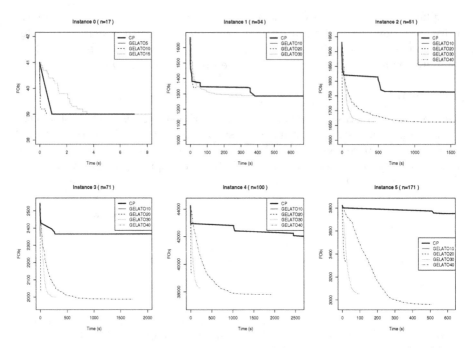

Fig. 3. Comparison between CP and GELATO (timeout for LNS is $0.25|FV|s$)

For the two small instance (instance 0 and 1), the two approaches are quite similar. In fact exploring a neighborhood of big dimension, very close to the whole search space (that with these instance is still tractable), is quite equivalent to solve the initial CP model. From instance 2 to the last, the trouble of CP in optimizing large search spaces comes out, and LNS gives better results.

The behavior of the CP algorithms is similar for all the instances: in the first seconds of the execution, CP finds some improving solutions, but from that moment it starts an exhaustive exploration of the search space without finding any other significant improving solution (the line of the objective function becomes quite horizontal). On the other hand, LNS algorithms have a more regular trend: the objective function is permanently improved during time, since a local minimum is found and then the algorithm is stopped (after 50 idle iterations).

Except for the LNS of 10 variables (too few), all the other LNS strategies find solutions improving the CP ones. They are also very competitive in time: some LNS algorithms overcome a lot the CP approach since the first seconds of the search (e.g. LNS with $|FV| = 20$ in instances 2,3,4,5, and LNS with $|FV| = 30$ in instances 3,4,5). This is due to the efficiency of LNS, that continuously explores different short parts of the search space, but in an exhaustive way, exploiting the constraint programming approach in any single neighborhood. On the other hand, the CP approach explores the whole search tree, but it remains bordered on a limited portion of it, so any improving solution outside this portion cannot be reached in acceptable time.

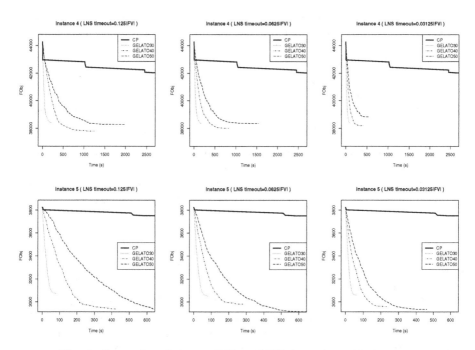

Fig. 4. Comparison between different LNS timeouts on big instances

Some comparison between the different LNS parameters (number of variables involved and timeout) can also be done: small LNS algorithms ("small" stands for LNS with few free variables) are very fast, and can make very big improvements in some seconds. Larger LNS algorithms are slower (they have bigger spaces to analyze), but they can reach better solutions: in fact they optimize a larger number of variables, so their exploration of the search space is more "global"; it could also happen that a neighborhood is too big to allow CP to perform an effective exploration. This trade-off must be investigates experimentally: e.g., from our tests it comes out that for instance 4, neighborhoods of size 40 are the best ones, while the ones of size 50 are too big, so ineffective.

The choice of the timeout for the exploration of a single large neighborhood is strictly connected to the dimension of the neighborhood: with small LNS (10, 20 or 30 variables) a long timeout seems to be better, because it can explore the whole (small) neighborhood and hopefully return the best neighbor. With bigger LNS (40,50 variables) it is the opposite: exploring the whole large neighborhood becomes a waste of time (at some point, constraint programming stops to find improvements), so it is better to stop earlier the search on the large neighborhood and continue quickly with another large neighborhood exploration.

A possible meta-heuristic that comes out from all these considerations could be the following:

1. start with small LNS with a short timeout(in this way we try to get the best improvement in the shortest time);

2. when no better solutions can be found, increase the timeout then launch Large Neighborhood Search (in this way an exploration has more time to explore the neighborhood and could fine better improvements);
3. iterate 2 since a big timeout value is reached;
4. increase the neighborhood's dimension, set the timeout the initial short value, then launch Large Neighborhood Search;
5. points 2–4 are iterated since the neighborhood's dimensions increases to intractable/ineffective ones.

This Multi-Neighborhood meta-heuristic should find large improving solutions in very short time (that can be quickly output to the user), and then run progressively deeper searches (even if more time expensive). Source codes and other technical details are in `http://tabu.diegm.uniud.it/EasyLocal++/`

5   Conclusions and Future Works

In this paper we showed that: using our GELATO hybrid framework, it is possible to combine a given CP model into a LS framework in a straightforward way; in particular, we can use a Gecode model as a base to define every kind of neighborhood, using the functionality provided by EasyLocal++ to execute competitive Large Neighborhood Search algorithms. We tested GELATO on several instances of the ATSP, and showed that performances of the hybrid LNS algorighms are very faster w.r.t. the pure CP approach, on all the non-trivial ATSP instances, even if the LS strategy is naive (Hill Climbing with random neighborhood selection). We also proposed a general Multi Neighborhood hybrid meta-heuristic that should improve the results we obtained so far.

We wish to extend the research pursued in this paper along two lines: developing and testing new hybrid algorithms; extending GELATO into a more general modeling framework. First of all we want to develop and test the Multi Neighborhood meta-heuristic proposed and other hybrid algorithms (e.g., based on Tabu Search, Steepest Descent, ...). We will implement these algorithms using GELATO and test them on a set of benchmark problems, more structured than the ATSP, also trying new problem-specific neighborhood definitions.

Concerning the second research line, we want to simplify the class hierarchy of GELATO (some classes and functionality must be added, other ones need to be cleaned up and refactored). Once GELATO has a more simple and clear interface, it will be integrated into a more general modeling tool to easily model and solve CSPs and COPs, that we already presented in [2]. In this tool, the user will be able to *model* a problem in a high-level language (e.g., Prolog, Minizinc, OPL), and specify the meta-heuristic he want to use to solve the problem; the model and the meta-algorithm defined will be automatically *compiled* into the solver languages, e.g. Gecode and EasyLocal++, since we use GELATO; at the end, the overall compiled program will be run and the various tools will interact in the way specified by the user in the modeling phase, to *solve* the instance of the problem modeled. We have already implemented two compilers from declarative languages to a low-level solvers (the Prolog-Gecode and

MiniZinc-Gecode compilers presented in [2]). We want to extend the functionalities of the compilers already realized and develop the *modeling-translating-solving* framework described above. Once this high-level modeling framework is well tested and reliable, developing hybrid algorithms will be flexible and straightforward and their executions will benefit from the use of low level efficient solvers.

References

1. Aarts, E., Lenstra, J.K. (eds.): Local Search in Combinatorial Optimization. John Wiley and Sons, Chichester (1997)
2. Cipriano, R., Dovier, A., Mauro, J.: Compiling and executing declarative modeling languages to gecode. In: Garcia de la Banda, M., Pontelli, E. (eds.) ICLP 2008. LNCS, vol. 5366, pp. 744–748. Springer, Heidelberg (2008)
3. Danna, E., Perron, L.: Structured vs. unstructured large neighborhood search. In: Rossi, F. (ed.) CP 2003. LNCS, vol. 2833, pp. 817–821. Springer, Heidelberg (2003)
4. Di Gaspero, L.: Local Search Techniques for Scheduling Problems: Algorithms and Software Tools. PhD thesis, Univ. di Udine, DIMI (2003)
5. Di Gaspero, L., Schaerf, A.: EasyLocal++: An object-oriented framework for flexible design of local search algorithms. Software — Practice & Experience 33(8), 733–765 (2003)
6. Focacci, F., Laburthe, F., Lodi, A.: Local search and constraint programming. In: Glover, F., Kochenberger, G. (eds.) Handbook of Metaheuristics, pp. 369–403. Kluwer, Dordrecht (2003)
7. Van Hentenryck, P., Michel, L.: Constraint-Based Local Search. MIT Press, Cambridge (2005)
8. Jussien, N., Lhomme, O.: Local search with constraint propagation and conflict-based heuristic. Artificial Intelligence 139(1), 21–45 (2002)
9. Monfroy, E., Saubion, F., Lambert, T.: On hybridization of local search and constraint propagation. In: Demoen, B., Lifschitz, V. (eds.) ICLP 2004. LNCS, vol. 3132, pp. 299–313. Springer, Heidelberg (2004)
10. Nethercote, N., Stuckey, P.J., Becket, R., Brand, S., Duck, G.J., Tack, G.R.: MiniZinc: Towards a standard CP modelling language. In: Bessière, C. (ed.) CP 2007. LNCS, vol. 4741, pp. 529–543. Springer, Heidelberg (2007)
11. Swedish Institute of Computer Science. Sicstus prolog, http://www.sics.se/isl/sicstuswww/site/index.html
12. Rossi, F., van Beek, P., Walsh, T.: Handbook of Constraint Programming (Foundations of Artificial Intelligence). Elsevier Science Inc., New York (2006)
13. Gecode Team. Gecode: Generic constraint development environment, http://www.gecode.org
14. Institut für Informatik Universität Heidelberg. Tsplib, http://www.iwr.uni-heidelberg.de/groups/comopt/software/TSPLIB95/
15. Various Authors. CP-AI-OR conference series, http://www.cpaior.org/
16. Wolpert, D.H., Macready, W.G.: No free lunch theorems for optimization. IEEE Transactions on Evolutionary Computation 1(1), 67–82 (1997)

Multi-neighborhood Local Search for the Patient Admission Problem

Sara Ceschia and Andrea Schaerf

DIEGM, University of Udine, Udine, Italy
{sara.ceschia,schaerf}@uniud.it

Abstract. We propose a multi-neighborhood local search procedure to solve a healthcare problem, known as the *Patient Admission* problem. We design and experiment different combinations of neighborhoods, showing that they have diverse effectiveness for different sets of weights of the cost components that constitute the objective function. We also compute some lower bounds on benchmark instances based on the relaxation of some constraints and the solution of a minimum-cost maximum-cardinality matching problem on a bipartite graph. The outcome is that our results compare favorably with the previous work on the problem, improving on all the available instances, and are also quite close to the computed lower bounds.

1   Introduction

Healthcare is surely one of the most important application domain of optimization in general and of metaheuristics in particular. Many papers have been devoted to healthcare problems, for example to nurse rostering problems [1], and more generally to timetabling of hospitals (see e.g. [2]).

The Patient Admission (PA) problem consists in assigning patients to beds in such a way to maximize both medical treatment effectiveness and patients' comfort. PA has been discussed in [3], formally defined in [4], and subsequently studied in [5]. In addition, Peter Demeester maintains a web site [6] publishing the available instances, the current best solutions, and also a solution validator to double-check researchers own solutions. The presence of the validator (a Java program, in this case) is very important, as it provides against the risk of misunderstanding in the formulation and in the cost components.

We propose a local search approach to the PA problem that makes use of different search spaces and neighborhood relations. We also study how to adapt the neighborhood relations for different weights of the components of the cost function. In addition, we propose a relaxation procedure to compute some high-quality lower bounds, which are useful to assess more objectively the quality of the solutions.

The outcome of our work is that our results are better than the ones obtained in [5], and also quite close to the lower bounds and to the optimal solution (for the only instance in which it is available).

M.J. Blesa et al. (Eds.): HM 2009, LNCS 5818, pp. 156–170, 2009.

2   Problem Definition

The general problem formulation is provided in details in [4]. We use a slightly simplified version, which is the one used in the public instances posted on [6]. We report this version here, in order to make the paper self-contained as much as possible.

2.1   Basic Formulation

These are the basic features of the problem:

Day: It is the unit of time and it is used to express the length of the planned stay of each patient in the hospital; the set of (consecutive) days included in the problem is called the *planning horizon.*

Patient: She/he is a person who needs some medical treatments, consequently must spend a period in the hospital and she/he should be placed in a bed in a room. Each patient has a *fixed* admission date and discharge date, within the planning horizon.

Bed/Room/Department: A room can be single or can have more beds. The number of beds in a room is called its *capacity* (typically one, two, or four). Patients may (with an extra charge) express preferences for the capacity of the room they will occupy. Each room belongs to a unique department. Patients in a room in the same day should be of the same gender.

Room Feature: Each room has different features (oxygen, telemetry, ...) necessary to treat particular pathologies. Every bed in a room has the same equipment. Patients may *need* or simply *desire* specific room features.

Specialism: Each patient needs a specific specialism for her/his treatment. Each department is qualified for the treatment of diseases of various specialisms, but at different level of expertise. In addition, each specific room has a set of specialisms it is particularly suitable for, each with its level of quality. Level are expressed in integer values from 1 (highest) to 3 (lowest).

The patient admission problem consists in assigning a bed to each patient in each day of her/his stay period. The assignment is subject to the following constraints and objectives:

Bed occupancy (BO): There can be at most one person per bed per day.

Room Gender (RG): For each day, patients in the same room should be of same gender; the gender of a room can change from day to day.

Department Specialism (DS): The department should have level 1 for the specialism of the patients hosted in the rooms of the department; lower level are penalized as explained in [6].

Room Specialism (RS): Similarly to departments, all levels lower than 1 are penalized (see [6]).

Room Features (RF): The room should have the features needed (or desired) by the patients, missing ones are penalized; the penalty is higher for needed ones than for desired ones.

Room Preference (RP): Patients should be assigned to rooms of the preferred capacity or with less beds.

Transfer (Tr): Patients should not change the bed during their stay; a bed change is called a *transfer*. Every transfer is penalized in equal way.

The constraint BO is obviously a *hard* constraint, in the sense that its violation makes the solution clearly infeasible. All the other constraints are considered *soft* in our formulation, and are assigned a weight w. In practice, the weights can be assigned by the user, based on the specific situation and policy.

2.2   Preprocessing

Based on the definitions given above, we recognize that the problem can be greatly simplified by means of two preprocessing steps which lead to a new problem formulation.

Room assignment. First of all, it is evident from the formulation that beds belonging to the same room are indistinguishable from each other in terms of features and constraints.

Therefore, we can reformulate the problem as an assignment of patients to *rooms* rather that to *beds*. To this aim, we have to replace BO with the constraint that the number of patients assigned to the same room in each day cannot exceed the capacity of the room.

Given that the output should be delivered in terms to assignments to beds, the room assignment must then be post-processed to be transformed to a bed assignment. In the solution delivered, a patient needs to be not moved from one bed to another one in the same room.

Ti this regard, it is easy to prove that if patients are ordered based on their admission day, and assigned to a fixed bed for the full stay, we produce an assignment that never moves a patient from one bed to another one in the room in consecutive days.

Patient-Room Penalty Matrix. The second preprocessing step is related to the notions of departments, specialisms, room features, and preferences. It is evident that all the constraints related to these notions contribute, with their weights, to the penalty of assigning a given patient to a given room.

Therefore, we can "merge" together this information into a single matrix that represents the cost of assigning a patient to a room. This *patient-room penalty matrix* C is computed once for all, when reading the input file, and all the four notions mentioned above (departments, specialisms, room features, and preferences) can be removed from the formulation.

Problem Reformulation. According to the first preprocessing step, the constraint BO is removed and replaced by the following one:

Room capacity (RC): The number of patients in a room per day cannot exceed its capacity.

Based on the second preprocessing step, the constraints DS, RS, RF, and RP are removed and replaced by the single following one.

Room patient cost (RPC): The penalty of assigning a patient p to a room r is equal to the value of $C_{p,r}$.

In conclusion, the problem includes one hard constraint, namely RC, and three cost components (or soft constraints): RG, RPC, and Tr.

It is worth noticing that, if we remove the constraint Tr, there is no correlation between the room assigned to a patient in one day with the one assigned to the same patient in a different day. In fact, in this case each single day could be scheduled independently.

3   Solution Technique

Our solution technique is based on a local search procedure. In order to describe it, we introduce the search space, the cost function, the initial solution, the neighborhood relations, and the metaheuristics.

3.1   Search Space

Differently from [4,5], our solvers search on the space, called S, of the patient-room assignments rather that on the space of the patient-bed ones.

The other difference w.r.t. the previous work is that we do not consider the space S of all possible assignments, but we restrict ourselves to some limited subsets of it. The first space that we investigate, called S_0, considers only states in which a patient is assigned to the same room for the full stay.

In all states in S_0 the cost of transfers is always 0, given that all solutions that contain transfers are removed by construction. Obviously, it is possible that we remove also all the optimal solutions.

For the above reason, we consider also the search space S_1, which is larger than S_0, but still smaller than S ($S_0 \subseteq S_1 \subseteq S$). In a state $s \in S_1$ we consider the possibility to transfer a patient *at most once* during her/his stay. Therefore, in a state $s \in S_1$ we can have *transferred* patients, that are moved in one day during their stay, and *unmoved* ones, that stay all the time in the same room. For transferred patients, we call *admission room* and *discharge room* the room of the first part and of the second part of the stay, respectively. A fragment of a state in S_0 and S_1 are shown in Figures 1 and 2, respectively.

The intuition behind the use of S_1 is that, in practice, it is quite unlikely that we have to transfer a patient twice, and in addition it is reasonable that the same quality of a solution with two transfers on a single patient can be obtained by a similar one with one transfer for two different patients.

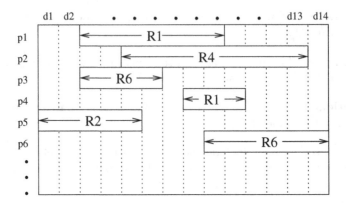

Fig. 1. The search space S_0 (d for days, p for patients and R for rooms)

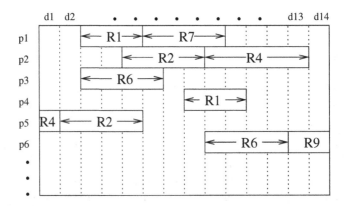

Fig. 2. The search space S_1

Summarizing, we have two distinct search spaces S_0 and S_1, that will be searched in different phases of the solution. For these spaces we will also use different neighborhood relations.

3.2 Cost Function and Initial Solution

The cost function obviously includes the cost components RG, RPC, and Tr. It also takes into account, with an appropriate high value, the constraint RC, given that assignments in which a room is occupied by more patients than its capacity are allowed.

The initial solution is generated in a random way: Assign to each patient a random room among all of them, independently of the capacity. However, a unique random room is assigned to each patient for the full stay length, so that in the initial state all patients are unmoved. This way the initial solution belongs to both S_0 and S_1.

Fig. 3. A CR move

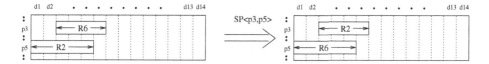

Fig. 4. A SP move

3.3   Neighborhood Relations

We present here several neighborhood relations, that will be used in combination as proposed in [7]. We first present the neighborhood relations for S_0. We consider the following two:

Change Room (CR): A patient is moved from one room to another one for her/his full stay. A CR move is identified by the pair $\langle p, r \rangle$, where p is the selected patient and r the new room.

Swap Patients (SP): The rooms of two patients are swapped for their full stay. A SP move is identified by the pair $\langle p_1, p_2 \rangle$, where p_1 and p_2 are two patients. The stay period of the two patients must overlap (for at least one day).

The reason why we limit swaps to patients with an overlapping is that otherwise the sets of the two changes would be totally independent. Thus the same movement could be obtained by two CR moves without the risk that the intermediate state is more "expensive" than the final one. Examples of CR and SP moves are shown in Figures 3 and 4.

Moving to the space S_1, we consider two other neighborhoods. The first one is an extension of CR that also introduces and removes transfers. Some examples of PCR moves are shown in Figure 5.

Partial Change Room (PCR):

  – For transferred patients: The admission room or the discharge room of a patient is changed to a new one. If the new room is the same of the other part of the stay, the transfer is canceled.
  – For unmoved patients: First it is decided if the move has to introduce a transfer or not:

  • If no transfer is to be introduced, then the move behaves like for CR.

- Otherwise, a transfer day, a new room, and the part (admission or discharge) are selected; the move then consists in inserting a transfer in the selected day and changing the admission or discharge room to the new one.

A PRC move is identified by the quadruple $\langle p, r, P, d \rangle$ where p is the patient, r the new room, P the part, and d the new transfer day. The attribute P can assume one of the three values H, T, F, which represent the first part of the stay (H for head), the second part (T for tail), or the complete stay (F for full). The attribute d represents the day of the insertion of the transfer, and it is meaningful only when both the following conditions hold: the patient is unmoved and $P \neq F$ (see second case of Figure 5).

Our next neighborhood is an extension of SP to deal with transferred patients. Examples of PSP moves are shown in Figure 6.

Partial Swap Patients (PSP): The rooms of the stay parts of two patients are swapped. For each single patient, if she/he is transferred, the stay part can be either the first or the second; if the patient is not transferred the stay part if the full stay. The transfer day is never modified. Like for SP, the swapped stay parts of the two patient must overlap.

A PSP move is identified by the quadruple $\langle p_1, p_2, P_1, P_2 \rangle$ where p_1 and p_2 are the patients, and P_1 and P_2 are the parts (H, T, or F). The part of a patient is necessarily F if the patient is unmoved, and H or T if she/he is transferred.

Notice that when a PSP move is applied to states in S_0, the attributes P_1 and P_2 have always the value F, and a move PSP behaves exactly like a SP move. Therefore, SP and PSP can actually be seen as a single neighborhood applied in different states. This is not the case for PCR; in fact, a PCR move in a state in S_0, would lead in many cases outside S_0 itself.

3.4   Metaheuristics

We have experimented with two metaheuristics, namely Simulated Annealing (SA) and Tabu Search (TS). Better results, in preliminary experiments, have been found by SA, and we therefore report here only the results for it. The version of SA used here is the standard one (see, e.g., [8,9]), with probabilistic acceptance and geometric cooling. The parameters are the start temperature T_0, the stop temperature T_{min}, the cooling rate β , and the number of neighbors N sampled at each temperature.

Given that different settings of the parameters of SA would result in different running times, in order to compare them in a fair experimental setting, we set a fixed timeout for all settings and we continue to *reheat* the solution until the timeout is expired. Reheating means that the temperature is set back to the start value after is has reached the minimum one, and the search starts again from the best solution found.

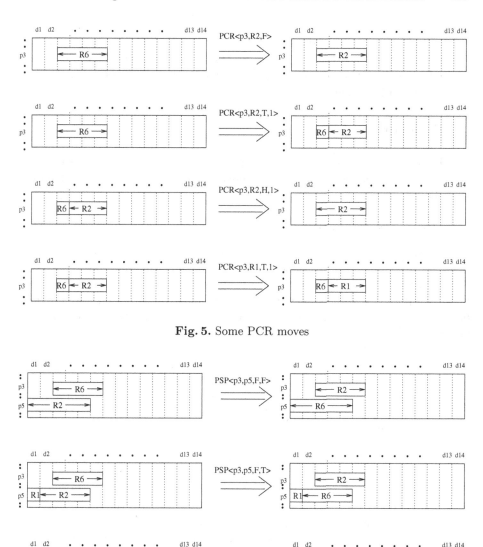

Fig. 5. Some PCR moves

Fig. 6. A PSP move

The first solver we consider, called M_0, is a SA using as neighborhood the union of CR and SP; using the terminology of [7] this would be defined as $SA(CR \oplus SP)$. It is clear that M_0 explores only the space S_0.

The second solver is a two-stage metaheuristic, that we call M_1. First it runs $SA(CR \oplus SP)$, in order to get to a good solution quickly, and subsequently it runs $SA(PCR \oplus PSP)$, starting from the best solution found by $SA(CR \oplus SP)$. In the terminology of [7], this would be defined as $SA(CR \oplus SP) \triangleright SA(PCR \oplus PSP)$, although the meaning of the symbol $\triangleright$ here is slightly different from [7], where the two solvers are used circularly. Here, this is not possible because CR can be applied only to S_0 states. For M_1 the available time is split between the two components in two equal parts.

4   Experimental Analysis

In this section, we first introduce the benchmark instances and the experimental settings, then we compute the lower bounds, and finally we show our experimental results.

4.1   Instances and Settings

We experiment on the same 6 instances used in [5] and available from [6]; their main features are shown in Table 1. For the number of patients we report the ones that are actually included in the problem. The one in parenthesis is the number reported in the instance files, but it includes patients that stay for zero days (same admission and discharge date) or have admission day after the end of the planning horizon.

The default values for the weights, defined in [6], are the following: $w_{RG} = 5$, $w_{DS} = w_{RS} = 1$, $w_{RFn} = 5$, $w_{RFd} = 2$, $w_{RP} = 0.8$, $w_{Tr} = 11$, where w_{RFn} and w_{RFd} are the weights for needed and desired room features, respectively.

Besides the default values, we investigate also on other combinations of weights for the cost components. Given that w_{Tr} is the most crucial one, we decide to experiment with many different values of it, while keeping all the others at the same level.

We set the timeout to the value of 200s for all experiments, on an Intel Quad-Core PC (64 bit) running Debian Linux 4.0. The software is written in C++, it uses the framework EASYLOCAL++ [10], and it is compiled using the GNU C/C++ compiler, v. 4.1.2. This is a much shorter time than the one granted in [5], which is 1 hour, on a Pentium D 3.4 GHz running Windows Xp.

All our results have been validated with the Java program available from [6].

4.2   Lower Bounds

We can make use of the patient-room penalty matrix C introduced in Section 2.2 to compute in polynomial time a lower bound of the cost for an instance. In fact, if we ignore constraints RG and Tr, the only cost is from RPC, i.e. the one related to C.

For each day, we can build a complete bipartite graph in which the left nodes represent the patients in residence in that day and the right nodes are the beds in the rooms (which instead are the same for all days). The weights of the edges are given by the elements of the matrix C, duplicated for each bed in the room. The problem reduces to computing the *minimum-cost maximum-cardinality matching* of this bipartite graph.

Table 1. Description of the instances

Instance	Number of beds	Number of rooms	Number of patients: actual (nominal)	Days in the planning horizon	Lower bound (LP-based)
1	286	98	652 (693)	14	636
2	465	151	755 (778)	14	1104
3	395	131	708 (757)	14	719.6
4	471	155	746 (782)	14	1074.2
5	325	102	587 (631)	14	618.4
6	313	104	685 (726)	14	769.6

It is well known [11] that computing the matching for this graph can be done to optimality using a linear programming (LP) formulation. We have therefore implemented an LP procedure using the public domain tool GLPK [12], and obtained an optimal integer solution of the matching problem (in a few seconds). The cost of this matching, summed up for all days, results in a lower bound for the available instances.

The lower bounds are shown in the last column of Table 1. As will be shown in Tables 3–6, these lower bounds are quite tight, showing that the costs related to the patient-room penalty matrix C are indeed the most significant ones.

4.3   Experiments with the Default Weights

Our first set of experiments is performed with the default values of the weights, which allow us to compare with the best known results available, also reported in the web-site.

Technically, in our solver we multiply all weights by 10 in order to have all integer values, and then use the faster integer arithmetic of the CPU. However, in the following tables we divide back the results by 10, so as to use the same scale.

In order to tune the solvers, we first apply the One-way ANOVA test to assess the significance of the SA parameters. We take into account the three numerical parameters T_0, T_{min}, and β, plus one additional non-numerical factor, called M in the table, that represents the solver and assumes the two values M_0 and M_1. The parameter N is fixed and set to the value 1000.

Table 2 shows the standard measures of the ANOVA test [13] for 30 runs on the configurations obtained by a full factorial design resulting from the use of two levels for each parameter: $T_0 = \{10, 100\}$, $T_{min} = \{0.001, 0.01\}$, and $\beta = \{0.9999, 0.99999\}$.

Looking at the p-values, which state the significance of the factor combination, we can conclude that the results are sensitive to all factors' combinations, excluding those with T_{min}. Given that results are insensitive to T_{min}, we set $T_{min} = 0.001$.

Table 3 reports the average results for the significant configurations for the 6 given instances. We can see that the best values (in bold) are obtained by different configurations on the various instances. Notice also the presence of

Table 2. Results of the One-way ANOVA test

	df	Mean Sq	F value	p-value
M	1	17.55	1055.74	0
T_0	1	12.50	752.03	0
T_{min}	1	$3.74 \cdot 10^{-7}$	$2.25 \cdot 10^{-5}$	0.9962
β	1	6.64	399.34	0
$M : T_0$	1	17.30	1040.50	0
$M : T_{min}$	1	$9.63 \cdot 10^{-5}$	0.005	0.9393
$T_0 : T_{min}$	1	$1.52 \cdot 10^{-7}$	$9.16 \cdot 10^{-6}$	0.9976
$M : \beta$	1	10.98	660.64	0
$T_0 : \beta$	1	7.81	469.76	0
$T_{min} : \beta$	1	$2.53 \cdot 10^{-4}$	0.01	0.9018
$M : T_0 : T_{min}$	1	0.001	0.031	0.8597
$M : T_0 : \beta$	1	11.19	673.18	0
$M : T_{min} : \beta$	1	$4.95 \cdot 10^{-5}$	0.003	0.9565
$T_0 : T_{min} : \beta$	1	$8.55 \cdot 10^{-5}$	0.005	0.9428
$M : T_0 : T_{min} : \beta$	1	$2.05 \cdot 10^{-6}$	0.0001	0.9911

Table 3. Results and pair-wise comparisons

	conf.	T_0	T_{min}	β	1	2	3	4	5	6	p-value
M_0	1	10	0.001	0.9999	677.1	1180.6	809.4	1235.3	634.2	829.0	0
M_0	2	10	0.001	0.99999	672.0	1163.1	802.1	1211.7	632.5	820.7	0.034
M_0	3	100	0.001	0.9999	**671.3**	1164.1	**795.8**	1207.2	633.9	818.2	—
M_0	4	100	0.001	0.99999	690.7	1165.9	801.4	**1206.9**	644.9	825.0	0
M_1	1	10	0.001	0.9999	676.5	1180.0	808.3	1235.7	634.4	831.4	0
M_1	2	10	0.001	0.99999	673.3	**1160.9**	801.1	1208.3	**631.8**	**819.2**	0.396
M_1	3	100	0.001	0.9999	673.3	1165.3	799.7	1213.9	633.8	820.4	0.002
M_1	4	100	0.001	0.99999	1261.3	1789.9	1422.1	1858.2	1035.2	1382.9	0
M_0	(avg. excluding				**673.47**	1169.27	**802.43**	**1218.07**	633.53	**822.63**	
M_1	configuration 4)				674.37	**1168.73**	803.03	1219.3	**633.33**	823.67	

very bad results in all instances for configuration 4 of M_1. These *outliers* are explained by the fact that for this setting the length of the run is excessive w.r.t. the time granted, so that the search is stopped when the temperature is still too high.

To select the "best" configuration to be used for comparison and further experiments, we apply the *Student's t-test*, and we report the corresponding p-value in Table 3. The p-value is meaningful if the underlying distributions can be assumed to be normal and independent [14]. If the calculated p-value is below the threshold chosen for statistical significance (typically $p < 0.05$), then the *null hypothesis*, which states that the two groups do not differ, is rejected in favor of an alternative hypothesis, which states that an algorithm is superior to another on the proposed instances.

To group together values from different instances, the cost is *normalized* w.r.t. the minimal value obtained for the instance. The dash in the column p-value identifies the configuration with the minimum average normalized cost. For the other configurations, this column shows the p-value of the comparison between the best configuration and the one of the current row.

Notice that $p = 0$ for four configurations, meaning that these can be considered inferior with high confidence. Only in two cases, we have $p > 0.05$, which denotes that these configurations are similar to the selected one.

The last two rows of Table 3 show the average cost for the two solvers, excluding configuration 4. The results highlight that there is almost no difference in performances between M_0 and M_1, but with M_0 slightly better that M_1. This means that for the default values of the weights, the exploration of S_0 is sufficient to obtain the best results.

4.4 Comparison with Best Known Results

Table 4 shows the comparison of the best configuration identified above (see Table 3) with the ones obtained by Bilgin *et al.* [5], reported in [6]. The timeout used in [5] is one hour, whereas we grant our solver 200 seconds.

The table also reports the percentage difference of the average w.r.t. our lower bounds shown in Table 1, and the best results obtained on all our experiments; finally, for instance 1 only, we report the optimal result which is available from the web site, that has been obtained by more that 60 hours of CPLEX computation (personal communication from P. Demeester).

Table 4. Comparison with previous work

Inst.	Our results				Bilgin *et al.*				Our overall best	Optimal solution
	Best	Avg.	(Dev)	ΔLB	Best	Avg.	(Dev)	ΔLB		
1	662.4	671.3	(4.28)	5.55%	750.4	801.48	35.35	26.02%	655.6	651.2
2	1149.4	1164.1	(6.41)	5.44%	1347.6	1383	27.65	25.27%	1143.6	—
3	777.8	795.8	(6.04)	10.59%	861	904.32	25.66	25.67%	773	—
4	1192.8	1207.2	(7.75)	12.38%	1450	1557.42	46.05	44.98%	1176.4	—
5	629.6	633.9	(2.13)	2.51%	648	655.44	6.1	5.99%	626.4	—
6	808	818.2	(4.75)	6.31%	910.4	932.06	14.1	21.11%	803	—

It is clear from Table 4 that we obtain better results than Bilgin *et al.* for all instances and in shorter computational time. In addition, our results are quite close to the lower bounds, and very close to the optimal cost, when available.

4.5 Experiments with Progressively Reduced Transfer Cost

To compare our solvers and to understand the relative effect of the constraint Tr, we test them also on a new setting in which w_{Tr} is progressively reduced from the default value down to 0. All the other weights are kept unchanged.

Table 5 shows the result of M_1 for different values of w_{Tr}. It also shows the average number of transfers in the solution (column #Tr). The horizontal line represents the threshold below which the use of transfers is profitable. This threshold depends on the specific instance, but it is included between $w_{Tr} = 1.3$ and $w_{Tr} = 0.6$ for most of them. Notice that some transfers are used also for values above the threshold, but without actual improvements of the overall cost.

Table 5. Results for experiments with progressive reduced transfer cost

Instances:	1		2		3		4		5		6	
w_{Tr}	Cost	#Tr	Cost	#Tr	Cost	#Tr	Cost	#Tr	Cost	#Tr	Cost	#Tr
11	672.5	0	1164.8	0	793.9	0	1208.7	0	634.0	0	820.6	0
5.5	672.0	0	1164.1	0	796.7	0	1209.2	0.4	634.2	0	819.4	0
2.7	672.4	0.3	1165.5	0.6	795.7	0.4	1205.1	3.6	634.3	0	818.7	0.5
1.3	673.3	4.0	1166.7	4.3	795.2	11.2	1200.8	20.6	634.8	1.3	818.5	5.4
0.6	670.9	13.6	1163.9	26.6	787.7	33.5	1184.6	57.4	634.3	5.7	815.6	22.4
0.3	666.3	23.4	1157.3	51.4	779.6	51.8	1174.9	85.0	632.8	20.3	811.4	41.5
0.1	661.7	36.7	1145.6	83.9	769.2	57.8	1158.6	114.5	627.7	28.7	801.3	49.1
0	655.8	227.4	1133.7	396.0	762.8	343.5	1146.5	377.9	623.2	321.9	793.2	326.1

4.6   Experiments with Day-Based Decomposition and No Transfer Cost

In order to further understand the landscape of the instances, we focus on the case that the cost of transfers is removed ($w_{Tr} = 0$). When transfers are not penalized, as already noticed in Section 2.2, each single day can be scheduled independently. Therefore, from a given instance we can create for each day belonging to the planning horizon a separated *sub-instance*, for which the planning horizon is exactly one day.

An example of a single sub-instance is shown in Figure 7. It is evident that the number of patients is reduced.

Therefore we solve with our solvers all the sub-instances separately, and we add up the costs to reconstruct the total cost. This way, we are actually exploring the original space S, without restrictions (thanks to the fact that Tr has cost zero). Given that all stays have length exactly 1, the search spaces S_1 and S_0 for the sub-instances actually coincide, so we use only the simpler solver M_0.

According to the fact that the problem is simplified, for this experiments we set the timeout to 100s in total for all sub-instances. The results are shown in Table 6, along with the lower bounds.

We see that the results are quite close to the lower bounds. The remaining distance to them is imputable to the constraint RG and to the non-optimality of local search.

For comparison, we report here also the results of M_1 for $w_{Tr} = 0$ (last row of previous table). It is evident that the results found here are clearly better than

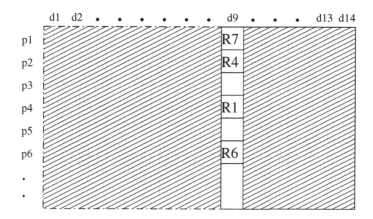

Fig. 7. The sub-instance for day d_9

Table 6. Results for experiments with day-based decomposition and no transfer costs

Instance	Decomposition				M_1 $(w_{Tr} = 0)$			
	Best	Avg.	(Dev)	ΔLB	Best	Avg.	(Dev)	ΔLB
1	638.4	641.39	(1.979)	0.85%	647.6	655.8	(5.667)	3.11%
2	1108.6	1116.51	(4.086)	1.13%	1121.4	1133.7	(6.024)	2.69%
3	736.8	747.05	(5.000)	3.81%	750.4	762.8	(9.034)	6.00%
4	1085.4	1101.02	(6.829)	2.50%	1119.8	1146.	(15.609)	6.68%
5	620	620.14	(0.314)	0.28%	620.8	623.2	(1.627)	0.78%
6	774.4	781.13	(3.514)	1.50%	781.4	793.2	(6.511)	3.07%

those of M_1. The explanation for this difference is twofold: First the restriction to S_1 is indeed a limitation for the case $w_{Tr} = 0$, second the navigation of a larger space is less effective than the solution by decomposition.

5 Conclusions

We have proposed both a simple local search solver M_0 and a composite one M_1 for the Patient Admission problem. Both solvers outperform previous work on the problem in terms of quality and computational time, despite the fact that they use the same technique, namely Simulated Annealing.

We believe that this difference in performance and speed can be attributed to both the preprocessing steps and the new neighborhood combinations.

We have also shown that the relative performance of the two depends, unsurprisingly, on the value of the weight of a crucial feature, namely the patient transfer.

Finally have shown that our results are relatively close to the lower bound provided by the linear programming solver for the underlying matching problem.

For the future, we plan to explore new combinations of neighborhoods and metaheuristics to obtained better results, and to experiment with new (possibly more challenging) instances.

Acknowledgment

We thank Peter Demeester and Greet Vanden Berghe for setting up the PA web site, for assisting us in the use of the validation, and for many fruitful discussions about the Patient Admission Problem.

We also thank Ruggero Bellio and Luca Di Gaspero for helping us in the use of the R package for the statistical analysis.

References

1. Burke, E.K., Causmaeker, P.D., Berghe, G.V., Landeghem, H.V.: The state of the art of nurse rostering. Journal of Scheduling 7, 441–499 (2004)
2. Cipriano, R., Di Gaspero, L., Dovier, A.: Hybrid approaches for rostering: A case study in the integration of constraint programming and local search. In: Almeida, F., Blesa Aguilera, M.J., Blum, C., Moreno Vega, J.M., Pérez Pérez, M., Roli, A., Sampels, M. (eds.) HM 2006. LNCS, vol. 4030, pp. 110–123. Springer, Heidelberg (2006)
3. Gemmel, P., Van Dierdonck, R.: Admission scheduling in acute care hospitals: does the practice fit with the theory? International Journal of Operations & Production Management 19(9), 863–878 (1999)
4. Demeester, P., De Causmaecker, P., Vanden Berghe, G.: Automatically assigning patients to beds resulting in improved hospital performance. Technical report, KaHo Sint-Lieven, Gent 06/2008 (2008)
5. Bilgin, B., Demeester, P., Vanden Berghe, G.: A hyperheuristic approach to the patient admission scheduling problem. Technical report, KaHo Sint-Lieven, Gent (2008)
6. Demeester, P.: Patient admission scheduling web site (2009), http://allserv.kahosl.be/~peter/pas/index.html (Viewed: June 1, 2009)
7. Di Gaspero, L., Schaerf, A.: Neighborhood portfolio approach for local search applied to timetabling problems. Journal of Mathematical Modeling and Algorithms 5(1), 65–89 (2006)
8. Aarts, E., Lenstra, J.K.: Local Search in Combinatorial Optimization. John Wiley & Sons, Chichester (1997)
9. Hoos, H.H., Stützle, T.: Stochastic Local Search – Foundations and Applications. Morgan Kaufmann Publishers, San Francisco (2005)
10. Di Gaspero, L., Schaerf, A.: EasyLocal++: An object-oriented framework for flexible design of local search algorithms. Software—Practice and Experience 33(8), 733–765 (2003)
11. West, D.B.: Introduction to graph theory, 2nd edn. Prentice-Hall, Englewood Cliffs (2001)
12. GLPK: GNU Linear Programming Kit, reference manual, Version 4.38 (May 2009)
13. Box, G.E.P., Hunter, J.S., Hunter, W.G.: Statistics for Experimenters: Design, Innovation, and Discovery, 2nd edn. Wiley Interscience, Hoboken (2005)
14. Venables, W.N., Ripley, B.D.: Modern applied statistics with S, 4th edn. Statistics and Computing. Springer, Heidelberg (2002)

Matheuristics: Optimization, Simulation and Control

Marco A. Boschetti[1], Vittorio Maniezzo[1], Matteo Roffilli[1], and Antonio Bolufé Röhler[2]

[1] University of Bologna, Bologna, Italy
[2] University of Habana, Habana, Cuba

1 Extended Abstract

Matheuristics are heuristic algorithms made by the interoperation of metaheuristics and mathematic programming (MP) techniques. An essential feature is the exploitation in some part of the algorithms of features derived from the mathematical model of the problems of interest, thus the definition "model-based metaheuristics" appearing in the title of some events of the conference series dedicated to matheuristics [1]. The topic has attracted the interest of a community of researchers, and this led to the publication of dedicated volumes and journal special issues, [13], [14], besides to dedicated tracks and sessions on wider scope conferences.

The increasing maturity of the area permits to outline some trends and possibilities offered by matheuristic approaches. A word of caution is needed before delving into the subject, because obviously the use of MP for solving optimization problems, albeit in a heuristic way, is much older and much more widespread than matheuristics. However, this is not the case for metaheuristics, and also the very idea of designing MP methods specifically for heuristic solution has innovative traits, when opposed to exact methods which turn into heuristics when enough computational resources are not available.

1.1 Optimization

Some approaches using MP combined with metaheuristics have begun to appear regularly in the matheuristics literature. This combination can go two-ways, either using MP to improve or design metaheuristics or using metaheuristics for improving known MP techniques, even though the first of these two directions is by far more studied.

When using MP embedded into metaheuristics, the main possibility appears to be improving local search (see [8] for a detailed overview). A seminal work in this direction is local branching [9], where MP is used to define a suitable neighborhood to be explored exactly by a MIP solver. Essentially, only a number of decision variables is left free and the neighborhood is composed by all possible value combination of these free variables.

The idea of an exact exploration of a possibly exponential size neighborhhood is at the heart of several other approaches. One of the best known is possibly Very

M.J. Blesa et al. (Eds.): HM 2009, LNCS 5818, pp. 171–177, 2009.

Large Neighborhood Search (VLNS, [3]). This method can be applied when it is possible to define the neighborhood exploration as a combinatorial optimization problem itself. In this case it could be possible to solve it efficiently, and it becomes possible the full exploration of exponential neighborhoods.

Complementary to this last is the *corridor method* [15], where a would-be large exponential neighborhood is kept of manageable size by adding exogenous constraint to the problem formulation, so that the feasible region is reduced to a "corridor" around the current solution.

Several other methods build around the idea of solving via MP the neighborhood exploration problem, they differ in the way the neighborhood is defined. For example, an unconventional way of defining it is proposed in the *dynasearch* method [7], where the neighborhood is defined by the *series of moves* which can be performed at each iteration, and dynamic programming is used to find the best sequence of simple moves to use at each iteration.

However, MP contributed to metaheuristics also along two other opposite lines: improving the effectiveness of well-established metaheuristics and providing the structural basis for designing new metaheuristics.

As for the first line, MP hybrids are reported for most known metaheuristics: tabu search, variable neighborhood search, ant colony optimization, simulated annealing, genetic algorithms, scatter search, etc. Particularly appealing appear to be genetic algorithms, for which a number of different proposals were published, with special reference to how to optimize the crossover operator. For example, Yagiura and Ibaraki [17] propose to fix the solution parts common to both parents and to optimize only on the remaining variables, while Aggarwal et al. [2] make a set consisting of the union of the components of the parent solutions and optimize within that set.

As for the second line, the proposals are different, but they still have to settle and show how they compare on a broader range of problems, other than those for which they were originally presented. One example is the so-called *Forward and Backward* (*F&B*) approach [4] which implements a memory-based look ahead strategy based on the past search history. The method iterates a partial exploration of the solution space by generating a sequence of enumerative trees of two types, called forward and backward trees, such that a partial solution of the forward tree has a bound on its completion cost derived from partial solutions of the backward tree, and vice-versa.

Another example uses classical decomposition strategies, namely Langrangean, Benders and Dantzig-Wolfe, to obtain metaheuristic algorithms, with a characteristic feature of evolving both a feasible solution and a bound to its cost [6].

1.2   Simulation

Simulation "is an attempt to model a real-life or hypothetical situation on a computer so that it can be studied to see how the system works. By changing variables, predictions may be made about the behaviour of the system" (source: Wikipedia). It is closely related to optimization, both because they often target the same applications in decision-making contexts and because prediction can often be made by optimizing a mathematical model.

It is with reference to this last possibility that matheuristics have been used in a simulation context. Optimization was used to determine how will system variables distribute under the assumption that the whole system leans toward a minimum free energy status.

For example, an application to traffic flow simulation was reported in [10]. In this case, the optimality assumption was formalized by the Wardrop's principles [16], which were taken as a guide for designing the objective function to be optimized by the matheuristic. The problem to solve could be stated as: given a road network with all needed parameters attached to the roads, an origin-destination matrix, and possibly traffic counters on some roads, determine the traffic flows on the roads. This can be formalized as a min cost multicommodity network flow (MCMNF) problem, with a nonlinear objective function, since the time for a driver to travel a road is a nonlinear function of the level of congestion of the road itself, as dictated by queuing theory. The cited application simulated the steady state in hours of maximum congestion and was applied to case studies relative both to cities (see figures 1 and 2 for an application to the case of La Habana, Cuba) and to whole regions, where the OD matrix counted thousands of rows and columns and the road network was composed of hundreds of thousands of arcs. In all cases the simulation time could be limited to a few minutes on a standard PC.

Figure 1 shows how congestion is correctly handled and, when paths followed by different drivers in congested situations are plotted, one can see that different choices were made for going from one same origin to one same destination. This ensures that no faster alternative way is available (Wardrop's first principle). Figure 2 shows the overall flow distribution for the 12 zones we considered in this study.

1.3   Process Control

Process control is "the task of planning and regulating a process, with the objective of performing it in an efficient, effective and consistent manner" (source:

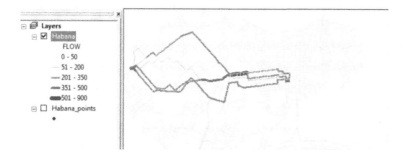

Fig. 1. Paths in Habana

Fig. 2. Flows in Habana

WikiPedia). Again, the request for efficiency and effectiveness calls for optimization, thus matheuristics can contribute to some applications.

Different use cases were reported. One refers to the identification of the best time-dependent bandwidth allocation parameters for peer-to-peer (p2p) nodes which should upload and download data [11], [5]. In this application, the proposed algorithm was a Lagrangean metaheuristics which could be fully distributed on the nodes of the peers themselves, allowing each node to optimize its own parameter in such a way that a global optimum would eventually emerge.

Figure 3 shows the simulated trajectory of the global network throughput, when nodes and connections were dynamically added or removed from the network. It can be seen how, after a brief startup where the nodes still had to decide who to connect with, a globally optimized state is quickly reached and fast adaptations to network structural variations could be ensured providing a

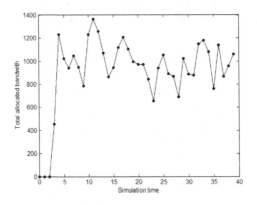

Fig. 3. Total system throughput

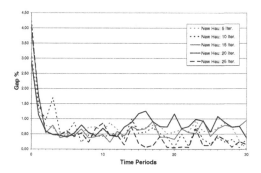

Fig. 4. Gap between the LP-relaxation and the heuristic solution

near optimal total system throughput. Figure 4 shows the evolution of the percentage gap between the LP-relaxation and the solution value provided by the Lagrangean metaheuristic for different parameter values. After the startup the gap is within the 1%.

Another application of matheuristics to process control aimed at determining the best distribution policy for a region-wide water network [12]. In that application, different infrastructural expansion of a current water supply network had to be compared in front of the needs and requests of different stakeholders, such as farmers, industries, civil population, water management authority, etc. A number of alternatives were identified and, for each of them, an optimal distribution policy had to be determined in order to assess the effectiveness of the particular infrastructure for each interest group. The distribution policies had to be able to cope with different environmental situations, such as years of drought followed by years of heavy rain.

The solution proposed made use of a *Receding Horizon Heuristic* (RHH), which is the metaheuristic counterpart of a well known process control technique named Model Predictive Control (MPC), or Receding Horizon Control (RHC). RHC is based on an iterative, finite horizon optimization of a model, and such is also the derived (meta)heuristic. At time t, the current system state is sampled and a cost minimizing control strategy is computed for a relatively short time horizon in the future: $[t, t + T]$. Only the first step of the control strategy is implemented, that relative to time t, then the system state is sampled again and the calculations are repeated starting from time $t+1$. This yields a new predicted state path which is implemented only in its first interval, that relative to time $t + 1$. Notice how the planning horizon keeps being shifted forward, hence the names RHC and RHH.

Computational results for this use case are plotted in figures 5 and 6. These show the level of service which could be guaranteed to the major town and the level of the biggest lake in the area, in a simulation considering 10 years of very different raining conditions (drought, heavy rain, than a run of drought years before returning to normality).

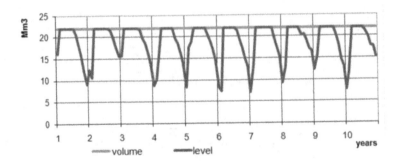

Fig. 5. Level of the reservoir

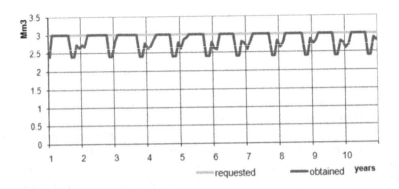

Fig. 6. City water requests

Figure 5 shows how the level of the reservoir, when compared with its usable capacity, decreases significantly during the summers, to recover during the winters. Figure 6 shows how the level of service to the main city, essentially making reference to civil usage such as drinking or washing, does never fall under an acceptable 80% of the request, which in turns will cause severe shortage for agricultural usage in drought years.

References

1. Matheuristics conferences page (2009),
   http://astarte.csr.unibo.it/matheuristics
2. Aggarwal, C., Orlin, J., Tai, R.: An optimized crossover for the maximum independent set. Operations Research 45, 226–234 (1997)
3. Ahuja, R., Ergun, O., Orlin, J., Punnen, A.: A survey of very large-scale neighborhood search techniques. Discrete Applied Mathematics 123(1-3), 75–102 (2002)
4. Bartolini, E., Mingozzi, A.: Algorithms for the non-bifurcated network design problem. J. of Heuristics 15(3), 259–281 (2009)
5. Boschetti, M., Jelasity, M., Maniezzo, V.: A fully distributed lagrangean metaheuristic for a p2p overlay network design problem. In: Hartl, R. (ed.) Proceedings MIC 2005 (2005)

6. Boschetti, M., Maniezzo, V.: Benders decomposition, lagrangean relaxation and metaheuristic design. J. of Heuristics 15(3), 283–312 (2009)

7. Congram, R., Potts, C., Van de Velde, S.: An iterated dynasearch algorithm for the single-machine total weighted tardiness scheduling problem. INFORMS Journal on Computing 14(1), 52–67 (2002)

8. Dumitrescu, I., Stützle, T.: Usages of exact algorithms to enhance stochastic local search algorithms. In: Maniezzo, V., Stützle, T., Voss, S. (eds.) Matheuristics: Hybridizing metaheuristics and mathematical programming, OR/CS Interfaces Series. Springer, Heidelberg (2009)

9. Fischetti, M., Lodi, A.: Local branching. Mathematical Programming B 98, 23–47 (2003)

10. Gabrielli, R., Guidazzi, A., Boschetti, M.A., Roffilli, M., Maniezzo, V.: Practical origin-destination traffic flow estimation. In: Proc. Third International Workshop on Freight Transportation and Logistics (Odysseus 2006), Altea, Spain (2006)

11. Maniezzo, V., Boschetti, M., Jelasity, M.: An ant approach to membership overlay design. In: Dorigo, M., Birattari, M., Blum, C., Gambardella, L.M., Mondada, F., Stützle, T. (eds.) ANTS 2004. LNCS, vol. 3172, pp. 37–48. Springer, Heidelberg (2004)

12. Maniezzo, V., Boschetti, M., Roffilli, M.: Matheuristics in simulation: a case study in water supply management. In: Caserta, M., Voss, S. (eds.) Proceedings MIC 2009, VIII Metaheuristic International Conference (2009)

13. Maniezzo, V., Stützle, T., Voss, S. (eds.): Matheuristics: Hybridizing Metaheuristics and Mathematical Programming. Annals of Information Systems, vol. 10. Springer, Heidelberg (2009)

14. Maniezzo, V., Voss, S., Hansen, P.: Special issue on mathematical contributions to metaheuristics editorial. Journal of Heuristics 15(3) (2009)

15. Sniedovich, M., Voss, S.: The corridor method: a dynamic programming inspired metaheuristic. Control and Cybernetics 35(3), 551–578 (2006)

16. Wardrop, J.G.: Some theoretical aspects of road traffic research, volume PART II, vol. 1. Institute of Civil Engineers, Palo Alto (1952)

17. Yagiura, M., Ibaraki, T.: The use of dynamic programming in genetic algorithms for permutation problems. European Journal of Operational Research 92, 387–401 (1996)

Author Index

Abdullah, Salwani   60
Arito, Franco   130

Blum, Christian   30
Bolufé Röhler, Antonio   171
Boschetti, Marco A.   171

Ceschia, Sara   156
Chaves, Antonio Augusto   1
Cipriano, Raffaele   141

Di Gaspero, Luca   141
Dovier, Agostino   141
Dubois-Lacoste, Jérémie   100

Ernst, Andreas   30

Hanafi, Saïd   73
Hussin, Mohamed Saifullah   115

Leguizamón, Guillermo   130
López-Ibáñez, Manuel   100
Lorena, Luiz Antonio Nogueira   1

Maniezzo, Vittorio   171
Mansi, Raïd   73
McCollum, Barry   60
Meyer, Bernd   30
Miralles, Cristobal   1

Pirkwieser, Sandro   45
Prodhon, Caroline   15

Raidl, Günther R.   45, 84
Roffilli, Matteo   171
Ruthmair, Mario   84

Schaerf, Andrea   156
Stützle, Thomas   100, 115

Thiruvady, Dhananjay   30
Turabieh, Hamza   60

Walla, Jakob   84
Wilbaut, Christophe   73